Center for Ethics & Humanities in the Life Sciences
C-208 East Fee Hall
Michigan State University
East Lansing, MI 48824

SUICIDE AND EUTHANASIA

PHILOSOPHY AND MEDICINE

Editors:

H. TRISTRAM ENGELHARDT, JR.

The Center for Ethics, Medicine and Public Issues
Baylor College of Medicine, Houston, Texas, U.S.A.

STUART F. SPICKER

School of Medicine, University of Connecticut Health Center,
Farmington, Connecticut, U.S.A.

VOLUME 35

SUICIDE AND EUTHANASIA

Historical and Contemporary Themes

Edited by

BARUCH A. BRODY

Center for Ethics, Medicine, and Public Issues,
Baylor College of Medicine, and
Department of Philosophy, Rice University,
Houston, Texas, U.S.A.

KLUWER ACADEMIC PUBLISHERS

DORDRECHT / BOSTON / LONDON

Library of Congress Cataloging in Publication Data

Suicide and euthanasia : historical and contemporary themes / edited by Baruch A. Brody.
p. cm. -- (Philosophy and medicine ; v. 35)
Includes bibliographies and index.
ISBN 0-7923-0106-4. -- ISBN 0-7923-0107-2 (pbk.)
1. Suicide--Moral and ethical aspects. 2. Euthanasia--Moral and ethical aspects. I. Brody, Baruch A. II. Series.
R726.S79 1989
179'.7--dc19 88-34141

ISBN 0-7923-0106-4

Published by Kluwer Academic Publishers,
P.O. Box 17, 3300 AA Dordrecht, The Netherlands.

Kluwer Academic Publishers incorporates
the publishing programmes of
D. Reidel, Martinus Nijhoff, Dr W. Junk and MTP Press.

Sold and distributed in the U.S.A. and Canada
by Kluwer Academic Publishers,
101 Philip Drive, Norwell, MA 02061, U.S.A.

In all other countries, sold and distributed
by Kluwer Academic Publishers Group,
P.O. Box 322, 3300 AH Dordrecht, The Netherlands.

All Rights Reserved
© 1989 by Kluwer Academic Publishers
No part of the material protected by this copyright notice may be reproduced or utilized in any form or by any means, electronic or mechanical, including photocopying, recording or by any information storage and retrieval system, without written permission from the copyright owner.

Printed in the Netherlands

TABLE OF CONTENTS

BARUCH A.BRODY

INTRODUCTION

One of the fundamental questions of contemporary bioethics is the treatment of terminally ill patients and patients (such as persistent vegetative patients) whose quality of life is very poor. Should such patients receive aggressive medical care or should they just be kept comfortable and allowed to die? Should they even be offered the option of assisted suicide or of euthanasia? What decisional process should be employed in making these decisions? Ever since the tragic case of Karen Ann Quinlan called public attention to these issues, the discussion of them has dominated the literature of bioethics.

Widely divergent positions have developed. They range from the view that each competent adult ought to have control over his life and ought therefore to be able to choose the time and manner of his death to the view that each moment of the biological existence of a member of our species is of infinite value and that physicians are therefore required to maintain life as long as that is possible. Many intermediate positions have been articulated as well.

A striking feature of the comtemporary debate is that it has paid little attention to the earlier discussions of these issues in the history of western thought. Unlike many questions in bioethics, the questions of suicide and euthanasia have been discussed at great length and with great sophistication throughout the history of western thought. Plato, Aristotle, and the Stoics in the ancient world, the medieval Jewish and Christian casuistrists, and such important modern philosophers and writers as Donne, Montaigne, Locke, Hume, and Kant all addressed these issues and had very important things to say about them.

The goal of this collection of essays – and of the conference sponsored by the Liberty Fund at which many of the papers were presented – is to make up for this neglect by carefully surveying the history of western thought on the questions of suicide and euthanasia and by reexamining the contemporary debate in light of that reexamination of the historical record.

I

The first three historical essays by Professors Cooper, Brody, and Amundsen attempt to correct certain widespread but mistaken images of the ancient and

Baruch A. Brody (ed.), Suicide and Euthanasia, pp. 1–7.
© 1989 *Kluwer Academic Publishers.*

medieval views on the questions of suicide and euthanasia. It is widely believed that the Ancients disagreed about the morality of suicide, with Plato and Aristotle categorically opposing it while the Stoics categorically approved of it. It is widely believed that Jewish casuistry unequivocally opposed all forms of suicide and euthanasia. It is widely believed that early Christianity had no objection to suicide, and may even have encouraged it by the admiration felt for martyrs, and that it was St. Augustine who first introduced the Christian objection to suicide. The essays in question challenge all of those beliefs.

Professor Cooper's essay carefully examines the views of Plato, of Aristotle, and of the Stoics, and shows that the usual picture of their views is much too simpleminded. The picture of Plato as an opponent of suicide is based upon his discussion in the *Phaedo*, and even there he at most tentatively accepts an account of the Pythagorean opposition to suicide. Elsewhere in his writings, however, he approves of certain types of suicide (in the *Laws*) and argues against aggressive medical care in certain cases (in the *Republic*). The picture of Aristotle as an opponent of suicide is based on Aristotle's claim that the person who commits suicide acts unjustly against the state; this claim is shown by Professor Cooper to involve confusions by Aristotle about his own views. The Stoics, he shows, were far from holding a categorical approval of suicide; the carefully formulated early Stoic view permitted suicide in certain cases (e.g., when beset by mental deterioration or by debilitating disease or by extreme poverty, or when necessary to discharge one's duties or to avoid disgrace) but opposed it in other cases (e.g., when one recognizes that one is likely to continue to do many immoral actions). In short, then, Professor Cooper's essay challenges many of the standard views about ancient thought on suicide. It also calls attention to the crucial ideas of the Neoplatonists Plotinus and Olympiodorus who, at the very end of the ancient world, actually returned to and defended the original Pythagorean position that suicide is never permitted.

Professor Brody's essay carefully examines the view that Jewish casuistry is committed to the belief in the sanctity of human life, the belief that each moment of biological life of every member of our species is of infinite value. While agreeing that traditional Jewish casuistry accepts the idea that suicide is prohibited as a form of self-killing and rejects the claim that one's life, body, and property are one's own to use as one sees fit, Professor Brody argues that these claims did not lead the Jewish casuistrists to accept the sanctity of life position. He shows that traditional casuistrists allowed certain forms of suicide (e.g., killing oneself out of fear that one will apostasize under torture, penitential and altruistic suicides, and suicides to avoid death by torture) and viewed the death of some patients as a blessing. As a result, some even prohibited providing care which prolongs the suffering of a dying patient and others allowed active pain relief even when that risked the patient's dying sooner. In short, Professor Brody's essay shows that the Jewish casuistrists supported a balanced approach among competing values rather than an absolute sanctity-of-life position.

Professor Amundsen's essay attacks the view that Augustine is the source of the Christian condemnation of suicide. His strategy involves his arguing for several different claims: (a) The New Testament, while never explicitly condemning suicide, does provide a structure of values and hopes inimical to suicide; (b) These values and hopes provided the foundation of the Patristic views on life and death, persecution and martyrdom, and sanctification and suffering. They led the pre-Augustinian Patristic authors to disapprove of suicide, even though they did approve of the virgins who killed themselves to avoid sexual defilement; (c) While Augustine may have been the first author to claim explicitly that suicide violates the sixth commandment against murder, his position is essentially that of the earlier Patristics except for his oposition to their approval of the virgins who committed suicide. This led Augustine to justify the suicides of those virgins who were treated as saints on the grounds that they acted under the influence of divine inspiration. Augustine also placed a great deal of emphasis on the virtue of endurance, but in this he followed many of the earlier Patristics.

The essays by Professors Brody and Amundsen provide a rich body of material to study the contrast between Jewish and Christian casuistry. Writers in both traditions discuss the relation between suicide and martyrdom, suicide and the avoidance of sin, and suicide and disgrace. They seem to come to conflicting conclusions on many of these issues. There is a need to better understand the nature of those differences and how they arose.

In any case, then, the first three of our historical essays lead to a rewriting of the early history of western thought about suicide and euthanasia. The next two essays, which assess the debate in the sixteenth through the eighteenth centuries, do not rewrite that history. What they do instead is to enrich our understanding of a history whose outline is already known.

II

The following summary represents what is relatively well known about the early modern discussion of suicide and euthanasia: Sir Thomas More and Michel de Montaigne both expressed some support for euthanasia and suicide. The first great modern defense of suicide was written, however, by John Donne in his famous *Biathanatos*. His leading successor was David Hume, whose essay "On Suicide" is the classic modern defense of suicide. Other important modern philosophers such as John Locke and Immanuel Kant opposed suicide. It is out of this dialectic that the modern debate begins.

The essays by Professors Ferngren and Beauchamp do not challenge this picture. What they attempt to do, instead, is to deepen our understanding of the argumentation employed in these early modern debates. If we are to learn from these debates, we will of course need to attend to the arguments used by these important thinkers.

Professor Ferngren carefully distinguishes the ideas of Sir Thomas More and of Michel de Montaigne. It is true that More, in his description of his Utopian society in his classic *Utopia*, describes the priests and magistrates as encouraging those who are suffering from painful and incurable diseases to put an end to their life or to allow others to release them. But it is hard to tell which portions of that book were intended by More as serious advice and which were intended to be just satire. The situation is very different with respect to Montaigne's essays on suicide. There, the discussion is removed entirely from its religious moorings, and Montaigne seems to have seriously advocated the view – basing himself upon many classic examples – that pain and the fear of a bad death are the best justifications of suicide. In this, he was followed by his friend and disciple, the priest Pierre Chavron, whose book on ethics was placed on the Index.

Both Professors Ferngren and Beauchamp carefully analyze the crucial book *Biathanatos* written by John Donne in 1606 or 1608 but not published until the mid-1640s (after his death) by his son. Although he does draw on some classical allusions, Donne's main strategy is to defend the morality of certain suicides against St. Thomas's claim that suicide violates the law of nature, the law of the community, and the law of God. Against the claim that suicide is unnatural, Donne argues that in some conditions people naturally desire to die. Against the claim that suicide harms the state, Donne argues that suicides would nevertheless be justified if the intention of the act were not self-promoting. Against the claim that it violates the law of God, Donne argues that none of the scriptural evidence backs that claim. He places particular emphasis on the fact that the suicides in Scripture are never condemned.

The evidence offered by both Professor Ferngren and Professor Beauchamp suggests that Donne's essay had little influence in his time. Jeremy Taylor's classic of Anglican casuistry, *The Rules of Conscience* (1660), reiterated the traditional arguments. John Locke's *Second Treatiste* (1690) carefully claimed that even in the state of nature, where one is free to dispose of one's personal possessions, one is not at liberty to destroy one's self. Perhaps the greatest influence was on such deist authors as Lord Herbert of Cherbury and his disciple Charles Blount.

Hume's writings on suicide, argues Professor Beauchamp, introduced some important new themes. His goal was not just to refute the traditional arguments against suicide offered by Aquinas and reiterated in the 18th century by Samuel Clarke. He also attempted to provide positive arguments for suicide. Professor Beauchamp claims that while Hume did appeal to the autonomy of individuals, his major argument was that some instances of self-caused death have good consequences. That is why he generally concluded that suicide is more justified when one's gain from dying is greater and the community's loss is less. Still, Beauchamp concedes, Hume sometimes appealed just to autonomy without regard to the social benefit or loss.

This issue of autonomy is also crucial to our understanding of Kant's position on suicide. Professor Beauchamp points out that some contemporary writers in bioethics mistakenly include Kant in their list of defenders of suicide because of his emphasis on individual autonomy. This, argues Professor Beauchamp, is clearly a mistake because Kant explicitly uses (in the development of his moral philosophy) suicide as one of his primary examples of a wrongful act. Kant's notion of autonomy is that of knowingly governing oneself in accordance with universally valid moral principles (ones which survive application of the categorical imperative) and not that of creating one's own values and one's own moral position. The relation between autonomy and suicide depends crucially on one's understanding of the concept of autonomy.

Professor Beauchamp calls attention to the fact that Kant, both in his early *Lectures on Ethics* and in his late *Metaphysic of Morals*, raises casuistrical questions about whether suicide for reasons other than self-love can be justified and as to whether it is permissible (or even obligatory) to allow oneself to die so as not to be dishonored. He feels that these hesitations suggest that Kant's position may not be as defensible as Kant himself thought.

III

The last two essays in this volume, the essays by Professors Boyle and Engelhardt, are rooted in this great historical debate we have been sketching and draw upon some of the themes developed in the course of that debate. Professor Boyle is particularly indebted to the Thomistic and Kantian theme of suicide as being in violation of human nature, while Professor Engelhardt draws on the themes of autonomy and individual freedom. Let us examine their arguments separately.

Professor Boyle begins by reminding us that Aquinas offered three different arguments against suicide, the three arguments to which Donne and Hume responded. Professor Boyle is interested in defending the first of the three arguments, the argument that suicide is wrong because it is contrary to the natural inclination of self-love. For Professor Boyle, this means that we learn from our natural inclination that human life is one of the fundamental goods and that any action which intentionally damages, harms, or impedes an instance of this fundamental good is wrong. It is this view that leads Professor Boyle to his moral condemnation of suicide.

Of special importance is Professor Boyle's account of the difference between his views and the well-known views of Alan Donagan. Donagan accepts the absolute prohibition against killing but does not give intentions the central moral significance which Professor Boyle attributes to them. This leads Donagan to adopt other strategies to allow as morally permissible those acts of killing which Professor Boyle allows as unintended killing. Professor Boyle in his essay suggests that Donagan's position might allow for too many cases in which evil is done for the sake of good.

All of this leads Professor Boyle, in the latter part of his essay, to consider the implications of his ideas for moral issues outside of the arena of bioethics. He is led, in particular, to challenge the moral permissibility of capital punishment and of the intentional killing of those engaged in criminal activity (as opposed to the unintended even if foreseen killing of them in the cases of self-defense). All of these consequences have, as he points out, considerable implications for our views about the prerogatives of the state in the use of force.

This last theme provides an excellent transition to Professor Engelhardt's essay, which is very much concerned about the legitimacy of the state's use of force. For Professor Engelhardt, the crucial question is the limit of state authority to restrict individuals who may wish to commit suicide or to provide to others the benefits of voluntary euthanasia, and not whether such actions are morally right or wrong. The latter, he claims, can only be established on the basis of a particular vision of the good life.

Why does Professor Engelhardt define the issues of suicide and euthanasia in the way in which he does, and why does he resolve them in favor of individual freedom? The answer to these questions lies in two crucial claims he advocates. The first is that we have to recognize the failure of mankind to establish on the basis of neutral rational arguments any particular conception of the good life as the canonical view of the good life. This puts him in opposition to Professor Boyle's version of Aquinas's theory of the basic human goods, of Kant's view about human life, etc. The second is his claim that the very definition of a moral community argues for a priority for individual freedom. He feels that a moral community is one which eschews the use of unconsented-to force against the innocent unless that use of force can be rationally justified in a neutral fashion, independent of one's conception of the good. All of this leads Professor Engelhardt to his defense of suicide and voluntary euthanasia.

In the latter part of his essay, Professor Engelhardt reviews the many moral arguments against suicide and euthanasia which have been offered in the history of western thought. He argues that none of them justifies a state prohibition either of suicide or of voluntary euthanasia. In this fashion, Professor Engelhardt's essay returns us to some of the themes we have been surveying in the essays in this volume.

JOHN M. COOPER

GREEK PHILOSOPHERS ON EUTHANASIA AND SUICIDE

The word "euthanasia" is not always used and understood in the same way. Nor, for that matter, is "suicide". The first thing to do in approaching the Greek philosophers' views about euthanasia and suicide is, therefore, to be clear about the senses of these words in which there existed for the philosophers of Greece corresponding moral categories. Which of the different kinds of action that might be called, or have by someone or other in modern discussions been called, euthanasia or suicide seemed to the Greek philosophers sufficiently interesting or problematic, from the moral point of view, so that they developed lines of argument and analysis in order to accommodate them? And which of these did they group together sufficiently closely for it to make sense to speak of Greek views on the morality of euthanasia or suicide? The answers to these questions will not just help us to avoid misunderstanding, by making it clear in what senses of these words it is acceptable to speak of Greek philosophers' views on euthanasia and suicide. They will also constitute an important first step in the substantial characterization of the Greek tradition in moral theory: one learns a lot about the character of any moral theory by seeing how, given that theory and its intellectual resources, the different kinds of human actions are arranged in significant groupings and which ones of these groupings are seen to call for philosophical comment.

I

Neither Greek nor Latin has a word that could be translated either "euthanasia" or "suicide". To be sure, our word "euthanasia" is borrowed from the Greek, but the Greek word (a coinage of the Hellenistic period) means simply a good death – an easy, painless, happy one or (possibly – so Cicero in *ad Atticum* 16, 7, 3) a fine and noble one. In fact, it was with the meaning of an easy, painless, happy death that the word "euthanasia" first entered English: the *O.E.D.* cites it in this sense as early as 1646 (1633 for the variant "euthanasy"), and it apparently continued to be used exclusively in that sense (with metaphorical extensions) until the 1860's. The earliest

Baruch A. Brody (ed.), Suicide and Euthanasia, pp. 9–38.
© 1989 *Kluwer Academic Publishers.*

citation the *O.E.D.* gives for "euthanasia" in the current (and now-dominant) sense of the action of inducing a gentle and easy death, especially as an act of mercy to those suffering from incurable and extremely painful diseases, is taken from W.E.H. Lecky, *A History of European Morals from Augustus to Charlemagne*, published in 1869.

Our word "suicide" is apparently a 17th-century formation on Latin roots, but classical Latin knows no such word. Classical authors have to resort to one or another of a set of noun- and verb-phrases that were in fairly standard use to refer to the act of intentionally killing oneself (likewise for the person who does it).[1] So does classical Greek, though there existed in late usage an adjective (and noun) *biaiothanatos* (or *biothanatos*) meaning "dying a violent death" that was applied especially to suicides in particular. Both in non-philosophical usage in Greek and Latin and (as we shall see) in philosophers' discussions of suicide, what gets counted as a suicide is always a death that a person both intended and brought about by some action of his own that was aimed, at least proximately, at bringing that death about. Cases, including some that might be described as "self-sacrificial", in which the agent knowingly risks death, even where the subjective probability of death amounts to virtual certainty, because he finds his own death, if that should in fact eventuate, an acceptable price to pay for the attainment of the goal being pursued, are not in Greek or Latin usage, or in the philosophers' discussions, grouped together with these intentional self-killings. So for our purposes these should not be described as suicides. This may seem (and is, in my opinion) natural enough not to call for special notice; but since in current discussions one sometimes finds the word "suicide" used, presumably under the influence, direct or indirect, of Durkheim, in such a way as to cover these willing self-sacrifices, as well as the cases where one's own death was itself actually intended, it is worthwhile making note explicitly of the restricted scope it is appropriate to give the word "suicide" in discussion of the classical philosophers' views. Hereafter, by "suicide" I mean a person's death both intended by him and brought about by some action of his own that was aimed, at least proximately, at bringing it about (or, of course, the person who brings about his death in this way).

II

There existed in Latin and Greek, then, standard ways of referring to suicide, so understood, and philosophers of all periods, beginning with fifth-century

Pythagoreans, had things to say about it. For euthanasia the case is different. Not only is there no single word in Greek or Latin that means (roughly) causing someone else's death in order to free him from an incurable, extremely painful or permanently debilitating disease or irreversible such condition; there is not even a standard phrase in general use, comparable to our "mercy-killing", having this meaning. Nor do any of the Greek systems of philosophical ethics seem to have selected precisely *this* kind of action for special consideration or comment. Probably this is partly to be explained by the fact that in antiquity people must have been fairly acutely aware of the uncertainty, given their current medical knowledge, of any judgment of incurability or imminent death, so that the conditions in which one might have found euthanasia a reasonable or even a mandatory course of action might have been so relatively rare that such actions either did not occur very often or, when they did, might have seemed aberrations that invited no special philosophical attention. No doubt religious prohibitions against the killing of human beings by private persons, backed by the threat of pollution and its ill effects on the killer, played a role, too.

But more needs to be said. For although there is, so far as I am aware, no discussion in an ancient philosophical text of the morality of killing, or allowing to die, persons who are incurably ill and *wish* (or may be presumed to wish) to die, Plato, in a well-known passage of the *Republic*, emphatically defends rules for the practice of medicine that would require some who are incurably ill not to be medically treated but instead allowed to die – without regard to their wishes (explicit or presumed) in the matter. In defending these rules Plato in effect applies a broader principle of social policy which selects as relevant features of the sick person's predicament not the pain caused by the disease, or the unpleasantness of the available medical treatment, or his own reflective assessment of the acceptability or supportability of continued life under the circumstances – the considerations we would look for in a case of euthanasia – but his inability, if treated and so kept alive, to continue to live the sort of full, active life devoted to socially useful employment that his nature and talents have previously suited him for. In so conceiving the issue, Plato places these persons and the question how they are to be dealt with in a broader category – one which includes, for example, defective new-borns, about whom parallel questions can be raised, and for whom, on fundamentally the same ground, Aristotle in the *Politics* defends similar treatment.[2] Thus, where Plato does approach most closely what for us would be potential candidates for euthanasia, his discussion makes it plain that he is not conceiving them so. This point is important enough to make it worthwhile to look closely at what Plato says about these cases.

In the part of the *Republic* in question (III, 405a-410a), Socrates is discussing what the practice of medicine will be like in the ideal city he is constructing together with Glaucon and Adeimantus. He insists that originally, with the first sons of Asclepius, who learned the practice of medicine directly from their father, the aim of medical treatment was limited to repairing damage due to wounds and ridding patients of "annual" diseases, i.e., maladies (especially, no doubt, infectious ones) that people are especially subject to at particular times of the year (405c8–9). Systemic disorders which if untreated would eventually lead to death but which 4th-century doctors could control so as to prolong the patient's life, but only by elaborate regimens involving special diets, special forms of exercise, prescribed periods of rest, etc., were not treated by the original Asclepiads. Socrates approves of this ancient scheme and adopts it for the practice of medicine in his ideal city.[3] His objection to such treatment of the systemic disorders is that it requires the patient to give himself over substantially and permanently to the management of his disease and so, in large measure, to give up the normal productive pursuits that characterized his prior life.[4] Such a person, he says (407e12, 408b12), would benefit (*lusitelein*) neither himself nor other people by his mode of life; the treatment would lengthen his life, but also make it a very bad one (*cf.* 407d6–7), and it is an abuse of the art of medicine to use it for that end. He should be allowed to die a natural death untreated.

Plato's policy here invites several comments. First, it clearly rests on the central contention of the *Republic*, that the sort of life that is in actual fact best for each individual person is one so organized that the good of other people in his community is significantly advanced by it. Being just, or, failing that, living justly, is a paramount good for any human being, and living justly requires living in such a way as to advance the good of others in the community. The patient Socrates describes has to abandon permanently those activities in which at once his own good in large part consists and the good of others is advanced, and that is what Socrates thinks justifies saying that such a person's life is of no benefit either to himself or to others. From the point of view of his own theory of the human good, Plato is not guilty of heartlessly requiring the death of people who are no longer, through no fault of their own, useful to society: the requirement is imposed equally, in fact primarily, for the good of the sick person himself.

But second, in accord with his conviction that what is good or bad for a person is an objective matter, to be determined by studying the facts of his case, the patient's wishes need not be consulted, and Socrates nowhere in the passage so much as mentions them. One who understands in what his good consists and has proper control over his desires will not want to have his life prolonged; for others, persuasion and, in any event, gentleness will be appropriate, but continued protestation will be unavailing. People who have fallen incurably ill in this sort of way will be allowed to die for their own good, whether they recognize their death as a good or not.

Third, it is important to notice that the treatments Socrates objects to immediately and necessarily deprive the patient not just of continued application to his previous productive life-pursuits but pretty well of *any* productive life at all. That is because, as Socrates describes these cases, the patient, in order to prolong his life, has to devote most of his time that might have been available for productive activity of one sort or another to staying alive: his time is simply preempted by the management of his disorder. The remedies are in this way immediately and necessarily self-defeating, if the purpose of a remedy is to restore to someone a life that he will be free to make some or other use of. This means that Socrates' policy would not have as a natural extension a rule that persons who have life-threatening accidents that one knows will result, even if treated, in their inability to return to their former work and other usual pursuits should not be ministered to.[5] In such a case the medical treatment would have a limited duration, after which the patient would be free, though with diminished capacities, to find something productive to do with his life, and even if it might seem quite certain, for one reason or another, that a given patient would not in fact succeed in finding anything useful to do, nothing in Socrates' remarks suggests that a doctor, or anyone else, should be qualified to opt for "euthanasia" in such a case.[6]

Fourth, it is perhaps worth adding that throughout his discussion Socrates seems to be considering only the treatment of persons in youth or mid-life. Elderly people who have lived past the time when active pursuits in the community's interest are in any event expected of them would not come under the provisions of Socrates' rule. For all he says here, doctors might be permitted to prolong the lives of elderly, retired persons, even by intrusive regimens that would not be permitted under the rule for people at other times of life. In Plato's republic, such persons, of whatever class, are due honor and respect, and retain their place in the community as members of the household (or its equivalent for guardians and rulers) valued for their past services and for their experience. The intrusive regimes that would undermine the lives of persons of other ages and social roles would not necessarily do so for these members of the community. Accordingly, the rationale Socrates uses to justify withholding medical treatment in the case of younger people would not apply to the very aged.

In this, as in other aspects of his social and political theory, Plato's concern is to arrange things so that people are really made better off, that is, are enabled and required to live so that they achieve what are in *fact* the best possible lives, the ones that are the best possible for them personally. Since what is good and what is bad for a person are objective questions, there is in general no reason to be guided by people's wishes in the application of social policies to them; a person's wishes are no reliable indication of what is in fact best for him overall. Hence in his prescriptions for the practice of medicine Plato makes no reference to consulting the wishes of the patient, but only to the actual, objective quality of his subsequent life if it should be prolonged. This means that what his rules provide for is euthanasia only in an extended, even somewhat Pickwickian, sense. Neither he nor any other Greek philosopher ever discusses euthanasia in our contemporary sense of the word.

III

With suicide things are quite different. From very early times Greek philosophers found occasion to discuss the rightness or wrongness, the appropriateness and rational acceptability, of suicide. In the *Phaedo*, that most Pythagorean of his dialogues, Plato has Socrates sympathetically report the views on suicide of Philolaus, an important fifth-century member of the Pythagorean brotherhood, about whom much was written in later antiquity but, unfortunately, not much was known (*Phdo.* 61b–62c). Diogenes Laertius, 8.85, reports that Philolaus wrote a single "book", and other evidence strongly implies that this was the earliest published writing by a Pythagorean; Pythagoras and Pythagoreans before Philolaus relied exclusively on secret, oral communication. However, Socrates' way of reporting Philolaus' views on suicide indicates that his knowledge of them was based on reports of what Philolaus had said in lectures or discussions (*cf.* 61d9), and not on anything in his book; and that the book contained nothing about suicide is confirmed by the fact that later Greek commentators on the *Phaedo* are unable to cite anything from what they knew as Philolaus' book to confirm or elaborate Socrates' report (see [2], pp. 223–229).

Interpreted narrowly, all Socrates says is that Philolaus maintained that it was not right to kill oneself (61c10, d6–7). He does not clearly attribute to Philolaus any reason for this prohibition; he goes on to suggest a reason that might actually support it, but this seems rather to be his own suggestion, presumably based on Pythagorean ideas but not reporting Philolaus' actual argument. It is noteworthy that Cebes, who admits to having heard Philolaus speak on this subject, is fairly emphatic that he has not heard from Philolaus or anyone else anything very clear about these matters (61d8, e8–9). But the rationale Socrates provides does cohere well with other Pythagorean ideas,[7] and it makes good sense of Socrates' reference to Philolaus to suppose that he means to be giving Philolaus', or anyhow a Pythagorean, justification of the prohibition. But whether the justification of the prohibition on suicide that Socrates proposes in the *Phaedo* is a pre-Platonic Pythagorean one, or merely Plato's own suggestion, its importance for later discussions of suicide can hardly be exaggerated. Socrates first (62b2-6) reports a theory (*logos*) he says is passed around in secret, as was typically reported of Pythagorean doctrines, that human beings live in a kind of "guard-post" (*phroura*), so that one must not (by committing suicide) "release oneself from (*heauton luein ek*) it and run away". It is not perfectly clear whether this guard-post is supposed to be a place where human beings serve as guards, keeping some kind of watch on behalf of the gods and under their direction, so that suicide is assimilated to desertion from one's battle-station, or a place, something like a prison, where we are kept under guard by the gods or their agents, in which case suicide would be compared to avoidance of some kind of judicial or quasi-judicial sentence. But Socrates' reference to the suicide's *releasing himself* from the guard-post strongly suggests that he is a captive there (Socrates might have spoken instead of abandoning one's post, if he had intended the other interpretation). And this seems confirmed when, just below, he says that, even if one does not accept this whole story, at least this much contained in it is reasonable, that the gods have charge of us (*einai hēmōn tous epimeloumenous*, 62b7): this seems a natural enough generalization from the idea of gods as the ones who are keeping us in prison, but much less natural if the idea was that they are our superior officers who command us in some further enterprise of their own. If so, the secret, presumably Pythagorean, theory has it that suicide is forbidden because it entails evading the full execution of a just sentence and so is itself something unjust.

Socrates, as just noted, is not willing to commit himself to this theory, apparently because of reservations he has about the idea that life is a punishment for something, but he does approve of the thought, which he finds expressed in it, that the gods are our keepers, who tend us and take care of us as possessions of theirs. As their possessions, we have no right to decide to cease to be tended and used by them as they see fit, any more than anything that belongs to us as a possession of ours has any right to decide not any longer to be our possession. Suicide, then, would be an injustice, a violation of the rights of ownership possessed by the gods in us. There is also the suggestion, to judge from Cebes' immediate response (62c-e), that suicide would be stupid, too, and wrong on that ground, because being under the charge and tendance of the gods, who are wise and good, we must expect to be exceeedingly well cared for in whatever way, as their possessions, we are treated, so that we can hardly do better for ourselves by committing suicide than the gods are doing for us in keeping us alive: when it *is* better for us to die, we can be sure the gods themselves, in their concern and tendance for us, will bring our deaths about.

In this whole passage, then, Socrates is offering to explain the Pythagorean prohibition on suicide, as put forward by Philolaus. The suggestion is that Philolaus did defend, or might plausibly, given other Pythagorean beliefs, have defended, this ban by arguing that human souls are placed in bodies by god as some kind of punishment, so that to commit suicide is to do the injustice of attempting to avoid serving the full term of a just sentence. Socrates himself sees in this theory, once questionable eschatological assumptions are pared away, the good idea that we are possessions of the gods and under their care and tendance. Hence suicide would be both an injustice (violating an owner's rights in his property) and the height of foolishness, since we should know that the gods, in tending us, will always act for our own good. If we do not die from causes the gods control, that can only be because it is better for us personally to continue to live than to die. Nonetheless, it deserves emphasis that this, though Socrates' own contribution to the discussion, is put forward simply as an explication and defense of the specifically Pythagorean ban on suicide, based on ideas Socrates finds plausible and attractive. He does not definitely commit himself to accepting the ban, even when defended in this plausible and attractive way. For he stresses in advance the tentativeness with which he entertains the idea, whether on these grounds or on others, that suicide is never justified. He prefaces his account of these reasons for the ban by saying (62a) that it would be strange that, when virtually everything else one can think of would sometimes, for someone, under some circumstances, be the proper thing to do, suicide alone should be plainly and simply *wrong*; surely, for some people it is better that they should die, and if so it is impossible to see why it should not be permitted to them to bring about their own deaths, rather than having to linger on until someone else does them the favor.

IV

The Socrates, then, of the *Phaedo* sees a good argument on each side of the question about suicide; he is no dogmatist on this question. It is true that in his account of the philosopher's attitude to death he seems to give more weight to the argument against it: he says (61c) that the philosopher will welcome his own death when it comes, as freeing him from dependence on the body and putting him finally into full contact with the ultimate truth of things, but will presumably not do himself violence since, as people say, that is not right (*ou themiton*). But this is stated fairly perfunctorily, and the argument, which I have just cited, that suicide must surely be sometimes permitted, is left standing, without any suggestion of rebuttal. This is significant because in the *Laws* Plato is quite explicit about the permissibility, and indeed the moral advisability, of suicide under certain circumstances. One does not find Plato elsewhere than in this passage of the *Phaedo* even seriously entertaining an absolute ban on suicide on general moral grounds.

In *Laws* IX, in discussing the criminal law, Plato twice has occasion to refer to suicide. The first comes at the very beginning of the book, in a kind of general preamble (854a3–5)[8] to what are presented as the most awful capital crimes (temple robbery, treason, political subversion, and others unspecified – it is not clear to me what the intended scope is of "similar crimes which are difficult or even impossible to cure", 854a12, and "all these impious deeds that bring about the ruin of the state", c6–7, which are the only indications Plato gives as to which crimes the preamble is meant to cover). The preamble is addressed to those who might be tempted to contemplate robbing a temple or committing some other such horrendous offense. First they are advised that anyone who feels any such temptation is subject to an evil impulse (an *epithumia kakē*, 854a6) that arises in human beings from no normal human or any divine origin, but as a result of unexpiated crimes done by other human beings (presumably their ancestors) in the distant past. One must do everything possible to rid oneself of, or at least control, such desires, destructive as they are both of the person's inner life who experiences them and of the social order. When any such thought enters your head, the lawgiver advises, you should seek relief by rites of purification and by supplicating the gods, and try to strengthen your own belief that every man has reason to honor what is fine and just (*ta kala kai ta dikaia*) by seeking the company of good people, listening to and trying to say, as your own conviction, what they say about this. If by this means the "disease" abates, well and good; but if not, "you should look upon death as the preferable alternative, and rid yourself of life" (854c4–5).

I have quoted from and summarized this passage so fully because it strikingly expresses a view about suicide – the view that suicide is justified when one's own moral character has proved irreparably to be very bad – that the Stoics later on strenuously denied. According to the Stoic theory, for reasons I will explain below, the moral goodness or badness of oneself and one's life are not only not important reasons against or for suicide, they do not count in the balance *at all*. Plato apparently thinks, to the contrary, that if one is subject to *extremely* immoral desires, which after serious efforts one can neither get rid of nor diminish so that they are fairly easily controlled, then one ought to end one's life. His thought, I take it, is not so much that suicide will preempt the possibility that at some future time one will yield to the desire and actually do something really horrendous, but that it will end a life that is so thoroughly bad, whether or not one does any of the horrendous things one is constantly wanting to do, that it is better for oneself not to live it at all.[9] Plato is thus, in this passage of the *Laws*, the most prominent opponent among their predecessors of the Stoics' doctrine that the moral quality of one's own life is in principle irrelevant to the question whether to go on living it.

Later in *Laws* IX (873c–d) the Athenian Stranger proposes a criminal law against (certain) suicides, and here too considerations having to do with one's own moral failings are mentioned as justifying suicide. What Plato says is this. With the exception of three special cases, suicides are to be punished with burial in unmarked, solitary graves in deserted, outlying districts. The special cases, for which neither this nor any other punitive action is to be taken, are these: when the agent acted (1) (as Socrates did) under judicial order, or (2) being forced (*anankastheis*, 873c6) by some excruciating and unavoidable misfortune, or (3) having come to participate in some irremediable disgrace that he cannot live with. Plato's language here, though it appears to be carefully chosen, is not completely clear to me,[10] but there seems to be a clear difference between the first two of these exceptions, on the one hand, and the third. The first and second, being cases where the agent is represented as acting under compulsion (*anankē*), legal or emotional – a typical case of the second kind might be suicide due to understandable grief or depression, caused perhaps by the loss of one's whole family in a fire – are apparently being conceived of as excused homicides. The third, however, appears to cover justified suicides (there is no reference in this case to compulsion), where the justification lies in the fact that the person has (perhaps intentionally, perhaps not – think of Oedipus!) done something *morally* very disgraceful: in the excellent city of Magnesia, for which these laws are being promulgated, one would not expect anything not involving moral failure to count as a disgrace, or at any event as a great enough disgrace to justify such drastic action. So, whereas earlier in book IX Plato had said that a person whose moral character was irremediably extremely bad should kill himself, here he counts suicide as justified as a way of extricating oneself from extreme moral disgrace brought on by one's actions.[11]

All other suicides than the three classes just indicated the law of Magnesia will punish in the way specified above, on the ground that anyone who commits suicide in other circumstances "imposes [an] unjust judgment [of death] on himself in a spirit of slothful and abject cowardice" (*argiāi kai anandrias deiliāi*, 873c7). Apparently, then, anyone who commits suicide when his judgment is unclouded by grief, depression, or other severely distorting emotions, because he considers that morally neutral bad things, such as pain, disease, the absence of interesting work to do, or the inability to do it, etc., so outweigh any good that his life can bring him that it is better not to go on living it, will be judged to have shown cowardice and a reprehensible unwillingness to take action against these evils and their effects on his life. That is to say, Plato here denies that the sorts of consideration that the Stoics later held *did* justify committing suicide ever actually do so, just as we have seen that he maintains that the moral considerations that they denied were ever even relevant to the decision sometimes in fact justified it. Plato's position in the *Laws*, therefore, appears to be, on both its positive and its negative sides, diametrically opposed to the Stoics'. As in other parts of their moral philosophy it seems reasonable to think of opposition to Plato's views as a principal component of the early Stoics' theories about suicide.[12]

V

But before turning to discuss the Stoics' views, I want to consider what Aristotle says about suicide in a short and difficult passage of the book on justice (*Nicomachean Ethics* V = *Eudemian Ethics* IV), 1138a5-14. The topic of suicide, which does not come up elsewhere in Aristotle's political and ethical writings, arises here in a discussion of the question whether it is possible for a person to treat himself unjustly, i.e., knowingly and willingly to do something to himself that is unjust. Suicide is knowingly and willingly killing oneself, and if to kill oneself knowingly and willingly is to do an injustice, then it seems to follow that, because one is oneself the victim of one's suicidal act, to commit suicide is to do *oneself* an injustice. Yet, as Aristotle sees, there is something paradoxical in the idea that it is possible to treat *oneself* unjustly. His aim in the passage is to address and attempt to remove this apparently paradoxical consequence of the existence of suicide, when this is juxtaposed with the fact of Athenian and Greek law generally, that suicide is legally forbidden.

In order not to misunderstand what Aristotle says here, it is important to bear in mind that on his theory of justice "just" and "unjust" can refer to either of two distinct sets of behaviors. On the one hand, they can refer to behavior characteristic of a person having a specific virtue or vice of character, one that has to do with the way he treats other people with respect to bodily harm and the distribution or assignment of external goods like money, property, etc. This specific virtue and vice are coordinate with the other virtues and vices Aristotle discusses in the central books of the *Ethics* – courage, temperance, good temper, etc. and their opposites, each with its own distinctive specific area of control over a person's behavior. On the other hand, "just" and "unjust" can also refer to behavior characteristic of someone who is simply law-abiding or lawless, who pays, or does not pay, heed to the law as such, who does or does not regard the fact that the law requires or forbids something as constituting some reason for or against doing it. In this broader usage what will count as just or unjust will depend directly on what the laws in fact do require or forbid. There is, antecedent to the institution of the law, no kind of action that counts as just or unjust in this broader sense, as of course there is for justice and injustice in the other, narrower sense, just as there is for courage and cowardice, temperance, self-indulgence, unreasonable self-denial, and so on.

Now this double usage of the terms "just" and "unjust" means that the question whether a person can treat himself unjustly can mean either of two things. First, can one knowingly and willingly treat oneself unjustly in the matter of bodily harm, assignment of external goods like money, and so on? Secondly, can one knowingly and willingly do something to oneself in violation of the law and so (in *that* sense) treat oneself unjustly? Since Aristotle thinks that, in general, the purpose of laws is to make people act in accordance with the whole range of the specific moral virtues he discusses, and so to help them to become morally good people, a well-framed legal system will include laws requiring the citizen to act justly (in the narrower sense), courageously, temperately, etc., and not to act unjustly, in cowardly fashion, and so on. Hence under such a system of law an act of injustice in the broader sense will also be an act of some specific vice as well – *perhaps* an act of injustice, forbidden by the law for the reason that that is what it is, but perhaps instead an act of cowardice, or one done from excessive or inappropriate anger, and so on.

Now Aristotle is careful to take note of the two types of justice in answering his question about the possibility of a person's knowingly and willingly doing something unjust to himself. It is in connection with injustice in the broader sense (and only in that sense) that he considers the case of suicide. The suicide does what the law forbids,[13] and, on Aristotle's theory, this means that in addition to doing something unjust (in the broad sense, i.e., illegal) he also does something contrary to one of the specific virtues. But which one? Aristotle does not indicate the full range of possibilities here; the (only) example he cites is one where the person acted in anger (a9–10), presumably a case where out of anger at himself for some real or imagined fault he kills himself, thinking that people with that fault don't deserve to live. So the case of suicide Aristotle considers is a case where the agent acted out of excessive anger; his was an act of the vice of irascibility, and it was as such that the law forbade it.[14] The question, then, whether in killing himself he has done anyone an injustice, and in particular whether he has done *himself* an injustice, ought to be simply the question who, if anyone, suffers the injustice that consists in his disobeying the law. Since, as described, the suicide is not an act of injustice in the narrow sense, but an act of the vice of irascibility, there should be no question of the suicide's doing an injustice to himself in the narrow sense. Understood that way the answer seems obvious, and it is the one that Aristotle himself goes on to give (a11–14): it is the city itself that has been unjustly treated by the law-breaker, viz., the suicide, and this is evidenced by the fact that what he suffers in punishment is some *atimia*, some loss of civic status (in the form, presumably, of an undignified burial: recall Plato's penalty in the *Laws*.)

In reaching this conclusion, however, Aristotle confusingly refers (a7–9) to the conditions in which, where someone harms someone else, he acts unjustly in doing so – conditions, that is to say, in which an act of injustice in the *narrow* sense takes place. This gives the, perhaps mistaken, impression that he considers the suicide done in anger also an act of injustice in the *narrow* sense (and so doubly, as it were, an act of broad injustice – a violation of the law against excessively angry behavior, and simultaneously a violation of the law against unjust behavior). And this may give his reader the impression that Aristotle holds that the suicide does an injustice in the *narrow* sense to his city – as if by depriving the city of his productive capacity or other services he takes away from his fellow-citizens a good that they, as a matter of justice, had a right to.

Now it is in fact easy enough to see how it might seem that the suicide-in-anger also does an injustice in the narrow sense to somebody. For to do an injustice in the narrow sense is simply to injure someone wrongfully with respect to bodily goods (health, life, etc.) or the distribution or assignment of external goods like money, etc. And of course the suicide clearly does at least do someone (viz., himself) an injury of one of these types. However, in order to be an act of injustice, his act must not merely do such an injury to someone. It must be a wrongful act of such injury, and wrongful in the sense that it is something forbidden by the particular virtue of justice. (That it is wrongful in the general sense of being forbidden by the law is, of course, not enough to make it wrongful in this particular way). And it is certainly not obvious that an act of suicide is a wrongful act in this sense; nor does Aristotle elsewhere say or imply that it is. I think Aristotle becomes confused at this point in his argument. Assuming, on the basis of existing law, that suicide is always unjust in the broad sense – i.e., always violates some particular virtue – and seeing that, of course, it always involves bodily injury, he unwisely and unnecessarily grants that suicide always involves *narrow* injustice (perhaps simultaneously with violations of other virtues as well). Pressed, then, with the question who is the victim of the injustice – who is the one who is unjustly treated by the suicide – he has to confront the common-sense intuition, an intuition he has himself defended less than two pages previously (1136b3–12), that one cannot do oneself an injustice in the narrow sense: injustice in the narrow sense requires two distinct persons as agent and victim. And so he concludes, confusedly, that not only is the city the victim of the suicide's illegal act – this makes good sense – but that the city, and not the suicide himself, is also the victim of a supposed injustice in the narrow sense that, in acting illegally, the suicide perpetrates.

Aristotle argues in the following way that the suicide-in-anger does his city an injustice (my comments are contained in parentheses):

1. When someone knowingly and willingly injures someone, contrary to the law and not in retaliation, he acts unjustly (– i.e., unjustly in the *narrow* sense).
2. But the suicide-in-anger slaughters himself knowingly and willingly, contrary to the law.
3. So he acts unjustly (– but with which kind of injustice? Injustice in the narrow sense, or in the broad sense? or in both?)

4. But he cannot be acting unjustly to *himself*, because he suffers knowingly and willingly, and we have already argued (1136b3–12 – but there the discussion concerned *narrow*-sense injustice only) that no one can knowingly and willingly be treated unjustly.
5. So the suicide-in-anger acts unjustly to his city, not to himself.

The problem with this line of argument is that at the beginning of the chapter Aristotle specifically introduces suicide as an action that naturally invites the question whether it is possible for a person to do himself an injustice in the *broad* sense. Aristotle has already completed his discussion of whether anyone can do himself an injustice in the narrow sense, giving an unequivocal answer of "No". Hence he ought to be discussing only the question whether it is possible for a person knowingly and willingly to do something illegal, and so unjust in the broad sense, of which he is himself the victim. So his conclusion at (3) ought to be that the suicide acts unjustly in the broad sense, and that the victim of this injustice – this illegality – is the city, his fellow-citizens in general. And it does follow from premise (2) that the suicide-in-anger acts unjustly in this sense; to establish that, it is sufficient to point out that he knowingly and willingly acts contrary to the law. By confusingly, and confusedly, adding premise (1) as well, Aristotle gives the impression that he means to say that the suicide acts unjustly in the narrow sense in addition. Furthermore, when he goes on in (4) to argue that the injustice done cannot be to himself, he *has* to have narrow-sense injustice in mind. For previously he has only argued that no one can knowingly and willingly be treated, or treat himself, unjustly in the narrow sense. So in this argument Aristotle does rely essentially on the highly questionable idea that in killing himself the suicide does a narrow-injustice.[15] Nonetheless, the conclusion he uses this erroneous idea to reach is, plainly, just that the suicide does his city, not himself, a *broad* injustice in killing himself, and that does not in the least entail that in doing so he violates the *rights* of his fellow-citizens (justice in the narrow sense), say rights they have to economic or other services from him. When Aristotle momentarily argues himself into suggesting this further point, it is in confusion, and is no part of the main line of argument he is pursuing.

VI

It is of some significance that Aristotle's and Plato's most extensive discussions of the morality of suicide come in the context of their treatments of the law. The Greek city-states generally had laws against suicide. At Athens, for example, suicides were buried with their right hands amputated – presumably in origin a measure designed to placate or render innocuous the ghost, as victim of a violent death, even though one at his own hands. And the thrust of both Plato's and Aristotle's remarks on suicide is to interpret in the light of their own general theories of justice and morality, and thereby to defend, at least the central provisions of the traditional legal codes in this matter. This reflects the conservative stance that both philosophers adopt toward the free city-state and its institutions (though this is perhaps nowadays more widely recognized in Aristotle's than Plato's case):[16] a major motive of their moral and political theories is to provide philosophically acceptable rationales for, if not the actual practices of any Greek city, at least a refined and purified version of the kind of social and political life that was led within Greek cities during classical times. No such motive was at work among the early Stoics. On the contrary, as the scanty reports about Zeno's *Republic* (*cf.* Plutarch *de Alex. virt.* I 6, SVF I 262; Clement and Plutarch in SVF I, 264), as well as the traditional associations of Zeno with the Cynic Crates, show, early Stoicism was sceptical, even iconoclastic, in its attitude to the provisions of traditional social and personal morality in the Greek city-states. And so one finds in the Stoic tradition, perhaps reaching as far back as Zeno himself,[17] a strong and very well-articulated defense of the appropriateness and correctness of suicide in many circumstances: whenever, for example, one can judge that, because of incurable illness, or extreme pain, or the absence of the necessary means to support oneself, a "natural" life, in the sense of a biologically smoothly functioning, unhindered one, is no longer possible. As I mentioned above, the Stoic position is the precise opposite of Plato's in the *Laws*: according to the Stoics, consideration of one's life prospects with respect to such external and bodily (so-called) goods does sometimes provide an adequate ground to put an end to one's life, but furthermore the fact that one is (and can expect to continue to be) a morally good person in itself gives no reason whatsoever in favor of continuing one's life, nor does the fact that one is a moral monster (and can expect to continue to be one), just in itself, give one any reason, however slight, in favor of putting an end to one's life.[18]

In holding this view the Stoics are simply applying to the special case of suicide their general theory of what a correct and appropriate action is in any given circumstances, and what the correct basis is for deciding what that is.

To explain their general theory fully and make it comprehensible would be a major undertaking in itself; I hope the following brief summary will be sufficient for present purposes. The Stoics, following up hints contained in arguments developed by Socrates in Plato's *Euthydemus* (278e–282d) and *Meno* (87d–89a), recognize two radically distinct kinds of value. First, there is the value (which they call goodness and badness) of that in which a person's good or ill (happiness or unhappiness, well-being or ill-being, etc.) actually consists. This they identify with a virtuous character and its exercise, and certain normal mental accompaniments of it, such as the joy, gladness, confidence, etc., that "supervene" upon virtue, and their vicious opposites. Anything other than your inner state and its expression in your decisions and actions is neither a good nor at all a bad thing for you; all else is indifferent, so far as being good or bad for you is concerned. But, second, there is the kind of value, subsidiary to this first kind, possessed by anything that, given its nature and the circumstances of human life, human beings generally have reason either to prefer to have or to avoid. It is right to prefer health to illness, for example; so a healthy life, though not better for you than a sickly one, is to be preferred: it has positive value of the second kind. The Stoics connected these two kinds of value to one another in the following way. A person's good (his being virtuous and acting virtuously) is the perfection of his nature as a rational being, that is, the condition in which his rational capacity is fully developed and properly employed, employed as nature intended reason to be employed in human beings. Now nature, as we can observe, sees to the growth, reproduction, maintenance, etc., and takes care, of the rest of the earth's animal life by means of the instincts and other non-rational impulses that such animals regularly experience and on which they act, in the light of the way things appear to them at time. For human beings, however, "reason has been bestowed ... for a more perfect way of leading their lives" (Diogenes Laertius 7.86) than nature provides in the case of the other animals. Reason is added in human beings, as Diogenes Laertius in reporting the Stoic theory puts it, as the "craftsman of impulse": the adult human being, like other animals, leads his life in response to his impulses, but in his case the impulses in question are themselves the product of his rational reflection upon and judgment about what is worth having and doing. The "more perfect way of leading their lives" that reason makes possible for human beings consists in the production by reason of the impulses on which we go on to act, and reason exists in us for the purpose of producing those impulses. This is where the second kind of value comes in. Our own reason, in taking on the direction of our lives, has to determine what sorts of impulses we ought to have, and directed to what ends. The things we ought to have impulses toward are just all the things that are (correctly) "preferred", and the things we ought to have aversions from are those that are (correctly) "avoided", i.e., the things that in fact have positive and negative value of the second kind.

But which are these? We can tell from observation of the course of nature that for all the other animal (and plant) species (leaving aside for the moment the human beings) there are certain norms for the members of those species that are in the general case (though obviously not in every individual case) attained in the life-span of its members. So we can see that it is part of nature's plan for the members of each species, in the general case even if not in every individual one, to grow to maturity in a certain way, perform certain sorts of activities including reproductive ones, and so flourish in a particular way that depends on the natural capacities that belong to members of that species. Plainly, then, for human beings, too, whose own reasoned impulses (and not *natural* impulses) direct their lives, those impulses will be correct that aim at the naturally flourishing life that corresponds for human beings to the flourishing life of a plant or a beast. In the human case this will involve, the Stoics think, being and remaining physically healthy; having sharp and unimpaired sensory and other organs; having certain sorts of family and other human relationships, including ones built upon mutual cooperation and help involving political and social ties giving rise to obligations to respect one another's independence and integrity; and having and using money and other material resources in furtherance of the objectives implied in all these natural human attachments and interests. These, then, are the norms that nature itself establishes for the conduct of a successful and flourishing human life, corresponding to that of a successful member of a plant or (brute) animal kind. They are "natural advantages" for a human being, and so are things that we have reason to take an interest in and try to obtain and maintain in our lives. Because reason is given to us by nature for a more perfect way of leading our lives than the other animals have got, we must conclude that we ought to shape our impulses so that we want to have and try to get, preserve and appropriately use, these things that contribute to the naturally flourishing life for a member of our species.

Thus, according to the Stoics moral virtue, the sole good for human beings, is, as it were, a purely formal condition: it consists in one's reason's being correctly informed about what things *other* than virtue itself are, by nature's plan for human beings, such as to promote the full and fully developed functioning of the natural capacities belonging to human beings as such, and shaping one's impulses to action in accordance with that knowledge. All the specific, substantive content of this state of mind – everything that determines what the virtuous person wants, cares about, makes an object of pursuit or avoidance in his actions, etc. – is drawn from the list of "preferred" and "avoided" (or "rejected") things, the things having value, positive or negative, of the second kind. Thus, Plutarch and Cicero both report the Stoics as maintaining that things of this kind constitute the "underlying material" for virtue, that which virtue judges about and chooses (or rejects) whenever it gets exercised in the virtuous person's life.[19] Thus, to pursue the good in which virtuous action consists is to pursue a purely formal end; one pursues it in pursuing some other, concrete goal, which is not and is not thought of as *good* at all (but only "preferred"), and in pursuing which, for that agent in those circumstances, virtue itself consists.

How does this general theory apply to the special case of suicide? Cicero states the Stoic view as follows (*de Finibus* III 60–61, trans. Rackham):

> When a man's circumstances contain a preponderance of things in accordance with nature, it is appropriate for him to remain alive; when he possesses or sees in prospect a majority of the contrary things, it is appropriate for him to depart from life... [T]he primary things of nature, whether favorable or the reverse, fall under the judgment and choice of the Wise Man, and form so to speak the subject-matter, the given material with which wisdom deals. Therefore the reasons both for remaining in life and for departing from it are to be measured entirely by the primary things of nature aforesaid. For the virtuous man is not necessarily retained in life by virtue, and also those who are devoid of virtue need not necessarily seek death... Even for the foolish, who are also miserable, it is appropriate to remain alive if they possess a predominance of those things which we pronounce to be in accordance with nature.[20]

It is easy to caricature the Stoic view: What's this? A person is *not* to take into account at all in deciding whether to continue his life or end it the only things that really matter for him, namely, what is for his own good and what is bad and harmful to him? Instead he is to decide this momentous question on the basis of how things stand for him that are literally neither good nor bad – things like his own continued health, the needs of his friends and family and so on, things which the Stoics count as relatively to be preferred, but, strictly speaking, absolutely indifferent so far as his own good or bad is concerned? Understandably enough, philosophers like Plutarch and Alexander of Aphrodisias, writing in the late first and second centuries A.D., when most philosophers were preoccupied with Plato and Aristotle, convinced of the superiority of their philosophies, and Stoicism had ceased to have able and original exponents, freely indulge in such caricatures and succeed in making the Stoic view seem ridiculous.[21]

But in fact their view is both coherent and not obviously implausible. First of all, it is not true that a fully virtuous person (a Wise Man) in deciding whether or not to commit suicide is not considering his own good (the continuance of his own virtuous inner state and its expression in action). He knows he will continue to be virtuous and will act virtuously (and so achieve good for himself) only if he does the appropriate thing, and in the right way, in the circumstances. Provided he does that, he will preserve and achieve his own good. So in deciding what *is* the appropriate thing to do, appeal to the continuance of his own good inner state and to continued good action on his part necessarily drops out: he will achieve those if, but only if, he does what he decides on other grounds, grounds having to do with non-moral values, is the correct and appropriate thing to do. And correspondingly, a morally bad person, concerned (as no doubt few such persons would in fact be) to avoid for himself the continuance of his morally bad inner state and its expression in action, must want to do the thing that is appropriate to his circumstances, and in the right way: that is the only way to avoid the harm that comes to oneself by being morally bad. Paradoxically, if the vicious person aims to rid himself of this burden by putting an end to his life, in circumstances where consideration of non-moral values alone would not support such an action, he only does one more vicious thing and so simply extends and confirms his possession of the bad. But, one might object, surely it is better for a bad person to do just this one more bad act, rather than to continue a life of repeated bad actions? Is not a shorter time in possession and use of the bad better, or anyhow less bad for him? And does not that give him a reason to commit suicide? The answer to these questions is: no, not in the least. Where a person's good and bad are at issue, only his moral state and its expression in action make any contribution. So it is no improvement in the goodness, or diminishment of the badness, of an agent's life to shorten the time he is morally bad; the only improvement in its goodness or diminishment in its badness there can be is for him to take steps to make a better person of himself. And the necessary first step in that direction is to start doing the appropriate things, as these are determined by the balance of non-moral reasons. But if these, taken by themselves, do not indicate that suicide is right and appropriate, then an agent who *is* truly concerned to improve his life or diminish its badness will refuse to commit suicide. Thus, as Cicero says, the bad person has *no* ground for appealing, in deciding whether to commit suicide, to anything other than the balance of reasons provided by the non-moral values that go to make up the class of things that are preferable and not preferable from his own point of view. It is not that a bad person has no reason to be concerned with his own moral state and with the fact that it is bad for himself to be like that; rather, a proper concern for that will lead him precisely to do whatever on these other grounds is rationally indicated.[22] And so, he will rationally commit suicide or not only because of how the prospects stand for him of a continued life in possession of a preponderance of "natural advantages".

VII

As one might expect, Epicurus, too, like the Stoics, did not hesitate to say that each of us is free to end his own life, if we encounter unendurable pain: at any rate, Cicero attributes to him the thought (*de Fin.* I 49) that we may "serenely quit life's theatre, when the play has ceased to please us", i.e., when it is causing us extreme pain that we cannot endure by recollecting previous pleasures and that we know will not be brief and intermittent. But, to judge from his extant remains, Epicurus was much more insistent on the *un*reasonableness of suicide than on its permissibility under such circumstances. He remarks in Vatican Saying 38 that, "he is of little account who finds many good reasons for departing from life", and Seneca (*Ep.* 24, 22-23) quotes three very interesting passages (frgs. 496, 497, 498 Usener) in which Epicurus analyzes as pathological the motives that lead many people to kill themselves. "It is absurd", he says, "to run towards death because you are tired of life, when it is by the manner of your life that you have brought it about that you ought to run towards death" – what you should do instead is to revise the way you live so that you no longer feel so tired of living that death is a reasonable option for you. Again: "What is so absurd as to seek death, when it is by the fear of death that you have unsettled and disturbed your life?" and, "So great is men's foolishness, indeed madness, that some are driven to death through the fear of death".[23] Epicurus diagnoses various deranged states of mind – especially, perhaps deep depression and acute anxiety – that cause some people to kill themselves as ultimately due to the irrational fear of death. Again, the right thing to do is to rid oneself of the fear of death, and so the state of mind that causes one to think of death as a reasonable option, rather than actually to commit suicide. So, although Epicurus did think suicide the right and appropriate thing under certain circumstances, namely, ones where the prospects for an acceptably pleasant subsequent life are irretrievably slight, he is insistent that many people find these prospects slight only because of the state of mind they have themselves fallen into and could get themselves out of if only they would listen to Epicurean reason.

VIII

The revival of Platonism as a system of dogmatic philosophy, beginning in earnest in the first century A.D., naturally carried with it a heightened interest in and appreciation for the views on the morality of suicide expressed in Plato's dialogues, and especially for those found in the *Phaedo*. Already by the time of Albinus, the leading Platonist teacher of the second century, the *Phaedo* occupied a central place, along with the *Alcibiades*, *Republic* and *Timaeus*, in the exposition of the Platonic system (see *Eisagoge* 5). Over the following several centuries a large number of commentaries on the *Phaedo* were written, in which the passage on suicide was a major focus of attention. Of the three surviving commentaries, the comments of Olympiodorus (sixth century A.D.) on the philosophical argument of this passage are particularly full and interesting.[24] In reviewing the Platonists' contribution to the debate about suicide we must also take into account the ennead (I 9) that Plotinus (third century) devoted to the topic, together with some relevant comments in the treatise on happiness (I 4), and a passage in the Neoplatonist Elias (sixth century) professedly reporting views of Plotinus apparently expressed in a passage that has not survived.[25]

For a Platonist like Plotinus the central question about suicide is whether it is ever recommended by, or even compatible with, the rational pursuit of the purification of one's soul, a purification that consists in the soul's finally realizing its inherent capacity to be completely engaged in thinking and understanding the system of first principles of reality (the Platonic Forms): it is in that final self-realization that happiness for a human being consists (see *Enneads* I, 4, 3, 24–40; I, 4, 4, 4–15). He seems to assume that Plato's view was that suicide never is justified in the pursuit of happiness,[26] and so sets himself to argue, especially of course against the Stoics, that when properly understood the true system of reality supports this opinion. Thus (the same is true, *mutatis mutandis*, of other parts of his work), Plotinus does not so much begin from an independent consideration of the question whether suicide is ever rationally justified in pursuit of one's own happiness, and the various arguments that might be developed on either side. He takes it for granted that the view of Plato, the wisest philosopher, who somehow infallibly knows the truth about all such fundamental matters, is the soundest and truest, though not always as fully and clearly argued for as one might wish. His own task in I 9 is simply to explain and defend the Platonic view, by supplying the arguments that are needed to reveal its truth fully to other philosophers.

One can distinguish five arguments in this very brief treatise (it totals only 19 lines), but the initial and central one (lines 1–8) is both highly original and specially interesting to us because it contains the basis of Plotinus' response to the Stoics. As I interpret it, this argument combines two lines of thought. First, it is a bad mistake to think that what one's soul needs, in order to exercise fully its capacity to think and understand the ultimate principles of reality, is to dissociate itself from the body in the sense of actually cutting itself free from it and going elsewhere. What is required is, rather, an internal transformation of the soul itself; the "return" to the first principle is not anything at all like a movement from one place to another (though that is a natural and useful metaphor for what is involved). In fact, the soul is always already separate from the body (I 9, line 5), i.e., a separate substance, and separation in the sense of a spatial removal is not in any event something that can happen to an immaterial entity like a soul. And in order for the needed transformation to take place, the attachment of the soul to the body is not in any way, or in any circumstance, a hindrance: that transformation depends entirely on the free exercise of a power of intellect that is in any case always entirely separate from the body. Just as God is always fully present to the material world, and the passing away of things within it depends entirely on their own unfitness to receive being from him (not upon any withdrawal on his part), so our soul should remain fully present to our body until our body itself through its own unfitness is no longer able to hold fast to it (I 9, lines 4–7, together with Elias, *Prolegomena* 6.15.23–16.2, which may be only a particularly full, expanded paraphrase of these lines).[27]

What enables Plotinus to argue in this way is his clear recognition that, although as he and the Stoics agree, human good consists in rational activity in accordance with the dictates of reason itself, there is a kind of rational activity that lies beyond and above the rational tendance of the body and living a correct social life. The Stoics think suicide is sometimes rational because they see as the only function of reason the maintenance of these lower life-activities; when they can no longer *be* maintained adequately, reason tells us to cease living altogether. But because, according to Plotinus, reason has a separate and prior task, a purely intellectual one (and one not at all aimed at tendance of the body or governing one's social relations), and this task is performed entirely without dependence on or reference to the body and its needs, or events that befall it,[28] one has no basis at all for arguing that reason itself (i.e., as it is in its own self) will ever dictate the termination of one's own life.[29] So, although the Stoics are right to insist that it is up to us whether to continue living or not, and "the door remains open", this choice will never be made by the wise and good person.[30]

To Plotinus' argument in defense of the Platonic thesis, Olympiodorus (Lecture I, section 2) adds two others. The first of these usefully expands upon Plotinus' claim that the pure rational activity of the intellect is independent of and not hindered by the soul's active tendance of the body.

"If God has two kinds of powers, elevative and providential, and if those by which he extends his providential care to secondary beings do not impede his powers of elevation and conversion upon himself, but he exercises both simultaneously, then there is no reason why the philosopher as God's imitator... should not be active creatively and providentially, while at the same time leading a life of purification..." (trans. Westerink).

The second, apparently expanding upon a remark at the end of the paraphrase in Elias, adds to the reasons why it is right to await our body's disengagement with our soul to bring about our death.

"The voluntary shackle should be unfastened voluntarily, the involuntary shackle involuntarily, and not conversely. That is to say, from natural life, which is involuntary, we should be released in the involuntary way, by natural death, while from a life in dependence upon passions, which we have chosen of our own free will, we should release ourselves in the voluntary way by purification" (trans. Westerink, with one change).[31]

IX

With the Neoplatonists Plotinus and Olympiodorus we come full circle, with a return to the Pythagorean view defended by Socrates in the *Phaedo*, that suicide is never permitted. I hope my exposition and discussion of the Greek philosophers' views on suicide have brought out the philosophical richness and interest of the treatment of the morality and rationality of suicide in this tradition. Practically all the recurrent theses in the subsequent centuries-long debate about suicide were not just adumbrated, but developed with such ingenuity and insight that one may with some justification feel that the Greek philosophers, taken collectively, already said everything of value on this topic.

NOTES

[1] For the action, Cicero writes "*mors voluntaria*", e.g., at *ad Familiares* 7, 3, 3, and *de Finibus* III 61, and "*mortem* (or *necem*) *sibi consciscere*", at *Brutus* 43; in discussing philosophical, especially Stoic, views about suicide he usually writes "*e vita excedere*" (or similar), translating the Greek circumlocution *heauton exagein ek tou biou* favored by Stoic writers in this context. Standard classical Latin phrases like "*vim* (or *manus*) *sibi* (or *suae vitae*) *adferre* (or *inferre*)" lay special emphasis on

suicides involving a violent attack on one's own life, by stabbing or by poison. Similar terminology is used in Greek.

[2] At *Pol.* VII, 1335b19–26, adopting the Spartan practice of having new-borns examined by magistrates to determine whether they are well-formed and fit enough to be allowed to live, Aristotle says that in the ideal city there should be a law that no deformed child (*pepērōmenon*) will be brought up. He gives no detailed justification for this law, but presumably it rests on basically the same ground as Plato's rules for the practice of medicine: allegedly, the congenitally deformed are (known to be) permanently debarred from developing the sort of full, active life in living which a human being's good consists.

[3] There seems no reason to doubt Socrates' seriousness in his prescriptions here; they follow quite clearly and directly from the overall theory of human good to which Socrates is committed in the *Republic*. Shorey's talk of "exaggeration" and "satire" ([13], p. 220) and his claim of the "humor" of the whole passage ([8], pp. 272, 276, n. *ad* 405d, 406e) should not mislead us, as they perhaps misled Shorey himself, into thinking that Socrates is only adopting a salutary pose here: there is (*pace* [8], p. 273 *ad* 406a) no discrepancy between Socrates' rejection of dietary regimens here and Plato's acceptance of them in *Timaeus* 89c, since the *Tim.* approves diet as the most effective means of *ridding* oneself of disease (*cf.* 89a5–6: diet is to be preferred to drugs as a means of purifying and restoring the constitution of the body), not as an acceptable way of prolonging a diseased existence.

[4] Socrates does not seem to recognize systemic disorders that could be managed effectively by less disruptive and intrusive regimens than those he objects to. But since his grounds for objecting to the disruptive and intrusive ones would not extend to prolonging the lives of sufferers from disorders that could be managed without forcing them to devote substantial amounts of their time and energy to the management of their diseases, Shorey is wrong to say that here Socrates austerely rejects "whatever goes beyond the training and care that will preserve the health of a normal body" (*ad* 407b). On the contrary, it is right to infer that he would have no objection to medical treatment for people whose diseases could be managed by not very intrusive and disruptive regimens: one thinks perhaps of insulin-and-diet treatment of diabetics.

[5] Note that at 405c8–9, in characterizing the ancient Asclepiads' practice of medicine, Socrates cites the treatment of wounds as one of the legitimate aims of medicine; there is no suggestion that a doctor would refuse to treat a treatable wound just because, in his judgment or in actual fact, the patient, though recovered, would not be able to lead a socially useful life.

[6] I use the word "euthanasia", in quotation marks, here advisedly. In standard contemporary usage the word is, I take it, applied only when the agent acts upon the wishes, express or presumed, of the patient to die, and for the sake of the patient's own good – not, as in the case envisaged (but rejected) here, partly because the patient has ceased to engage in any employment useful to the community. It is worth noting however that H.J. Rose (a classical scholar) actually defines euthanasia in such a way that the future usefulness of a person's life would be a prominent consideration in acts of euthanasia: "Euthanasia may be defined as the doctrine or theory that in certain circumstances, when, owing to disease, senility, or the like, a person's life has permanently ceased to be either agreeable or useful, the sufferer should be painlessly killed, either by himself or by another" [12]. Was Rose, self-consciously or not,

showing the influence of this discussion in Plato, or of (what he took to be) Greek attitudes? (It was, perhaps, knowledge of Greek that led him here also to classify at least some suicides as special cases of euthanasia.)

7 See especially Iamblichus, *Vita Pythagori* 86 (Diels-Kranz 58C, vol. I, p. 465.5–6): one ought to beget children, for it is our duty to leave behind us other people to worship the gods – a piece of Pythagorean oral teaching apparently adopted and adapted by Plato in *Laws* VI, 773e. On this see [3], p. 171. Reference is also sometimes made (e.g. by Rose, *art. cit.*) to the Pythagorean idea that a soul's embodiment is a punishment by God for sins it committed in a past life, so that suicide would be an offense against justice (as well, of course, as perfectly futile, since it would be an additional sin for which an extension of the period of embodiment would be the expected penalty). It is true that Philolaus is quoted (by Clement of Alexandria, *Strom.* III 17 = Diels-Kranz 44 frag. 14) as having written that souls are buried in bodies as punishment, and that Athenaeus attributes to one "Euxitheos the Pythagorean" the thought that because that is so suicide is wrong and will call down further punishment on the offender; but this reasoning is not attested for Philolaus. Whether one finds some such line of thought in or lying behind the *Phaedo* passage will depend on how one interprets Socrates' reference at 62b to the "guard-post" in which human beings are supposedly placed: is this a post in which we are *under* guard (in effect, therefore, in a kind of prison), or one where we are *serving* as guards (so Cicero interpreted it, *cf. de Senectute* 73)?

8 In citing the *Laws* I use Trevor J. Saunders' translation.

9 It is perhaps worth reminding oneself at this point that in the *Republic*, for example, Socrates is emphatic that what really matters for a person, and so the ultimate source of all one's real reasons for acting, is the good internal condition of one's own soul (see, e.g., 443c9–444a2). Doing bad things is wrong because it brings about or reinforces or simply expresses a bad internal condition of the soul. So, too, here in the *Laws*, Plato is not saying that the potential criminal's suicide is ultimately desirable so as to prevent him or her from doing harm to society: doing that harm is itself to be avoided, on Plato's view, only because it is the expression or the cause of a bad internal psychical condition, and it is the avoidance or the riddance of that that must provide the fundamental reason for the suicide.

10 It is not clear, for example, why it matters that the misfortune under (2) should have been *unavoidable*.

11 Thus it seems correct to say that in 873c–d Plato exempts from punishment suicides on moral grounds (viz., as responses to extreme moral disgrace) that are closely connected to those he earlier recommended (854c4–5) – the suicides of those incurably afflicted with morally awful desires. But it is an exaggeration to say (with Apelt, nn. to 854c and 873c in his translation [7]; and see [1], p. 163) that the exemption *explicitly* covers these recommended cases. There is no mention at all in the earlier context of overt criminal acts, giving rise to disgrace, which is clearly what the *aischunē* in 873c refers to. So the suicides recommended earlier would have to be understood as covered by the exemption here granted, if at all, only by a natural extension of the provisions of the law as explicitly formulated.

12 In my discussion of Plato's views on euthanasia and suicide I have limited myself to discussing views that are put forward and discussed in the context of a philosophical argument, i.e., in the context of what Plato himself describes as *logos* in contradistinction to *muthos* (e.g., *Protag.* 320c3–4, 324d5–6). As is well known, Plato appends

eschatological myths to illustrate and extend the philosophical content developed in the *logos* of certain of his dialogues. Especially in the cases of *Gorgias, Phaedo* and *Republic*, these myths have much to say that is relevant to the topic of suicide. However, as Socrates' discussion of his own rhetorical displays in the myths of the *Phaedrus* makes abundantly clear, myths convey truth in a derivative and secondary way and are not to be confused with properly philosophical exposition. (Nor is it in the least a straightforward matter to decide how they are to be interpreted – as Socrates also makes clear). Accordingly, in discussing Plato's philosophical views one must begin from the philosophical argument, taken on its own terms, and if one goes on to consider the myths, one must always control their interpretation by reference to the philosophical argument they are intended to illustrate. Where the myths go beyond anything established by argument, their contents cannot be attributed to the character Socrates, much less to the author Plato, as items of philosophical opinion.

[13] The manuscript reading at 1138a6–7 (translated by Ross) makes Aristotle adopt the very peculiar view (not elsewhere reported) that what the law does not require of us it forbids us to do, so that since no law requires (non-judicial) suicide the law forbids it. J.A. Stewart ([14], p. 533) attempts to defend this line of thought, by emphasizing that "law" (*nomos*) really means customary at least as much as statute-law (it is more plausible to say that what custom does not require, it forbids), but not very successfully: there are plenty of matters about which custom is silent, not requiring but also not forbidding specific ways of acting (e.g., drinking iced water with meals). Aristotle himself seems to speak of the law as directly forbidding suicide just one sentence below (*ouk eāi*, a10), and goes on to speak of an established penalty for breaking the law against suicide (a13), which seems to imply that he is thinking of a statute expressly prohibiting it (as in Plato's *Laws* IX), so he does not need to rely on this dubious line of thought to reach a legal prohibition of suicide. There is much to recommend Joachim's emendation of the text at 1138a6–7 (translated by Irwin), which makes Aristotle say simply that the law forbids anyone to kill himself.

[14] Presumably, like Plato in *Laws* IX, Aristotle thinks that other suicides will be motivated by excessive fear, and so will be acts of cowardice rather than irascibility. The crucial point is that, if he thinks the law will contain a blanket prohibition of suicide, he must think that *every* suicide will be brought about by some excess or deficiency that makes the act an act of some or other specific vice. So he thinks that anyone who *thinks* he has adequate reason to kill himself only thinks that because of the distorting influence on his process of reasoning of some desire he either ought not to have had, or ought to have been able to resist the influence of.

[15] What is worse, Aristotle's own account of narrow-injustice appears to imply that this idea is actually erroneous. If the claim in steps (1)–(3) is that the suicide does a narrow-injustice *simply* in that he knowingly and willingly injures someone (viz. himself), then this conflicts with Aristotle's theory. According to this account of when someone does a narrow-injustice (1136a31–b5), there must be someone who is injured (the one who is treated unjustly) and this injury must be something his *boulēsis* (his rational desire) is actively opposing at the time. But the suicide does not have a rational desire *not* to die, at least not when he acts, so *he* is not unjustly treated. And it certainly does not seem that the city's *boulēsis* (that of the fellow citizens in general) can be actively opposing the suicide, since there is no reason to suppose anyone at all even knows it is taking place. That means that simply *in* killing himself he does not do

an act of narrow-injustice at all. At 1138a7–9 Aristotle misstates his own account of when a person acts unjustly, by omitting the requirement that what is done is contrary to the rational desire of the victim.

[16] But Hegel in the *Philosophy of Right* is rather emphatic in saying that Plato's *Republic*, at any rate, is not at all a mere philosopher's ideal, unconnected except negatively with Greek politicial and social reality at Plato's time: it is actually nothing but "an interpretation of the nature of Greek ethical life" ([5], p. 10; see also paragraph 185, p. 124). For Plato, no less than for Hegel himself, the owl of Minerva spreads its wings only with the falling of dusk.

[17] Seneca reports (but without citation, and in a form that may suggest that "Zeno" here just means "the Stoics") that whereas Socrates will teach you to die if it is necessary, Zeno will teach you to die before it is necessary (*Ep.* 104, 21).

[18] Consideration of what, being a good person or a moral monster, one will *do* for or to other people if one remains alive is of course another matter; as we shall see, on Stoic principles that counts as relevant to the decision whether to kill oneself or not.

[19] Plutarch, *de Communibus Notitiis* 1071b; Cicero *de Finibus* III 61. See also Arius Didymus in Stobaeus *Eclogae* II 47.12–48.5 Wachsmuth, on *hypotelis* vs. *telos* in virtuous action.

[20] In this passage Cicero is thinking exclusively of the potential suicide's private good – his health, his ability to carry on an active life. Other sources make it clear that other concerns, e.g. concern for his friends or for his country (Diogenes Laertius 7.130), could appropriately motivate suicide: the good of one's friends and one's country are also among the things that are naturally "preferred" by human beings. Olympiodorus, the 6th century A.D. Neo-platonist commentator, reports ([15], *Comm. on Plato's "Phaedo"*, I, 8, 19–39) that the Stoics recognized five cases where suicide is appropriate (so also Elias, *Eisagoge* in *CAG* XVIII 1, 14.15–15.22, cited in SVF III 768): (1) in discharge of some duty, e.g., to defend one's country; (2) to avoid doing something disgraceful, e.g., betraying an important secret when pressed by a tyrant to do so; (3) when beset by mental deterioration in old age or (4) incurable, debilitating disease; (5) when extreme poverty prevents one from supplying one's basic needs.

[21] See Plutarch, *de Stoicorum Repugnantiis* ch. 18, *de Communibus Notitiis* ch. 11, Alexander of Aphrodisias *De anima* II, pp. 159.16–22, 160.20–31, 168.1–20 Bruns.

[22] The Stoic view as expounded in Cicero and discussed above is that of the original Stoics of the 3rd century B.C. Later Stoic writers, such as Seneca and Epictetus have a good deal to say about suicide, and maintain the old Stoic position that suicide is sometimes appropriate (see Seneca, *Ep.* 14, 70, 77; Epictetus, *Discourses*, I, 9, 10–17; I, 24, 20; I, 25, 18; II, I, 19–20; etc.). But Seneca tends to forget that according to Stoic doctrine virtue *requires* a full commitment to life, so long as it continues, as well as permitting (even requiring) departure from it under certain circumstances. He writes as if, really, life itself is a burden and a bore, so that one is free to leave it *whenever one pleases* (see the last lines of *Ep.* 77: "stop living whenever you *want*"!) Epictetus, too, though he goes out of his way to deny that one should return promptly to God (*Disc.* I, 9, 10–17), is responsible for encouraging that view in the negative and world-weary way he describes our lives in "this" world. The most he can say on the other side is that God has stationed us in our bodies, so we should wait for a signal from him to leave them (a reflection, of course, of Socrates' discussion in Plato's *Phaedo*) – a far cry from the carefully articulated, anything but pessimistic and world-weary, theory of the old Stoics.

[23] Compare the remarkable passage of Lucretius, III 79–84, in which he recounts how fear of death drives people to all kinds of foul deeds, in an effort to accumulate wealth and other resources as a bulwark against death – including finally the foul deed of their own suicide.
[24] Olympiodorus' commentary is printed with English translation and notes in [15].
[25] This passage is printed, for example, as an appendix to I 9 in [10]. For the different scholarly views about its relation to I 9, see Armstrong's introduction thereto ([10], pp. 320–321).
[26] I take it he bases this opinion on the *Phaedo*; what he made of the *Laws* passages he nowhere makes clear, but perhaps he believed the *Laws* meant to recommend suicide only for the sake of the good of others, and not as a step towards one's own fuller achievement of happiness.
[27] Combined with this first line of thought is the idea that if one does commit suicide, instead of waiting for a natural demise (or some other death, by accident or the action of another soul), the involvement of one's own soul in actively separating the body from its hold on the soul will leave the soul itself still somehow bound up with the body – the separation will only in any case be complete if it is the *body* that loosens its grip on the soul, not the other way about.
[28] On this see *Enneads* I 4, 2, 38–55; 4, 18–32.
[29] As my exposition of Plotinus' argument in I 9 and I 4 makes clear, I do not accept A.H. Armstrong's view (footnotes to pp. 192 and 324–25) that Plotinus, in either treatise, regards suicide as reasonable under any circumstances. Plotinus is explicit at I 9, lines 11–14 (and compare I 4, 9) that incipient madness is no reason to commit suicide – and what could Plotinus, with his views about the value of rational activity, think *would* give us reason to commit suicide, if going mad did not? And in I 4, 8, 1–2 and 8–12 he makes it clear that if pains become unbearable without ceasing of their own accord, they (not oneself) will bring one's life to an end. Though he pointedly says (lines 8–9) that under such circumstances one retains one's power of choosing what action to take in the face of such torments, he plainly excludes the choice of suicide: as he goes on to say (lines 24–30) the good and wise man will set his virtue against these and all other adversities, thus keeping his soul unperturbed by them. Armstrong, in holding that despite all this, Plotinus admitted suicide as legitimate in "absolutely desperate circumstances" (similarly R. Harder, [4], p. 546), apparently has in mind (or could have in mind) three passages (I 9, lines 15–17; I 4, 4, 31–2 and 43–5), but I do not think any of them need be so interpreted. The reference to some necessity's leading to one's death in advance of the destined time (I 9, 15–17) need not envisage anything more than a death caused not by natural events but by another person's action (or, perhaps, the reference may be to the sort of judicial and forced suicide Socrates underwent – Plotinus' *anankaion* may recall Socrates' *prin anankēn tina theos epipempsēi, Phaedo* 62c6, as the *phamen* at I 9, 17 perhaps indicates). The two passages of I 4, 7 say no more than that it is up to oneself what to do in response to being taken into slavery, and that one is free to depart, by committing suicide, *if* under those circumstances it is not possible to be happy (*eudaimonein*) – and, as we have seen, *his* view is that under those circumstances that definitely is *not* impossible. It is important to emphasize that none of the interpreters who, on the basis of these passages, take Plotinus to have thought suicide rationally justified under certain circumstances has yet explained how this opinion is to be derived from, or even squared with, his theory of human *eudaimonia*.

[30] I think J.M. Rist is probably right ([11], pp. 174–77) to point to Plotinus' acceptance of the Stoic thesis of our freedom to commit suicide at I 4, 7, 44–5 (and see lines 31–2 and I 4, 8, 8–9), while combining it with the claim that it is never right to *use* this freedom that way, as a pointed rejection of the Stoic view about the reasonableness of suicide under some conditions.

[31] Oddly, Olympiodorus cites (without explanation) Plotinus' treatise I 9 (under the title "On Reasonable Suicide") as providing reason to think suicide is sometimes permitted (1.8. 17–18). I have no explanation for how he can think this. And, when he gets round to drawing his own conclusions about suicide (1.9) he surprisingly says that suicide *is* sometimes permitted, whenever it is beneficial to the soul – but he does not say when that would be, or respond to Plotinus' arguments to show it never could be beneficial to the soul.

BIBLIOGRAPHY

[1] Apelt, O.: 1912, *Platonische Aufsaetze,* B.G. Teubner, Leipzig.
[2] Arnim, J. von: 1903, *Stoicorum Veterum Fragmenta* (= SVF), B.G. Teubner, Leipzig.
[3] Burkert, W.: 1972, *Lore and Science in Ancient Pythagoreanism,* Harvard University Press, Cambridge.
[4] Harder, R.: 1956, *Plotins Schriften,* vol. I (b), Felix Meiner, Hamburg.
[5] Hegel, G.W.F.: 1952, *Hegel's Philosophy of Right,* trans. by T.M. Knox, Clarendon Press, Oxford.
[6] Lecky, W.E.H.: 1869, *A History of European Morals from Augustus to Charlemagne,* Longmans, Green, London.
[7] Plato: 1916, *Die Gesetze,* trans. by O. Apelt, Felix Meiner, Leipzig.
[8] Plato: 1937, *The Republic,* Books I–V, trans. by P. Shorey, Harvard University Press (Loeb Classical Library), Cambridge.
[9] Plato: 1970, *The Laws,* trans. by T. Saunders, Penguin, London.
[10] Plotinus: 1966, *Enneads 1–2,* vol. 1, trans. by A.H. Armstrong, Harvard University Press, (Loeb Classical Library), Cambridge.
[11] Rist, J.M.: 1967, *Plotinus: The Road to Reality,* Cambridge University Press, Cambridge.
[12] Rose, H.J.: 1914, 'Euthanasia', in J. Hastings (ed.), *Encyclopedia of Religion and Ethics,* T&T Clarke, Edinburgh, vol. 5, 598–601.
[13] Shorey, P.: 1933, *What Plato Said,* University of Chicago Press, Chicago.
[14] Stewart, J.A.: 1892, *Notes on the Nicomachean Ethics of Aristotle,* vol. I, Clarendon Press, Oxford.
[15] Westerink, L.G.: 1976, *The Greek Commentaries on Plato's Phaedo,* vol. I, North-Holland Publishing Company, Amsterdam.

Princeton University,
Princeton, New Jersey,
U.S.A.

BARUCH A. BRODY

A HISTORICAL INTRODUCTION TO JEWISH CASUISTRY ON SUICIDE AND EUTHANASIA

The belief in the sanctity of human life is the belief that each moment of biological life of every member of our species is of infinite value. This belief has profound implications for discussions of medical ethics; it stands in opposition, for example, to most recent discussions of death with dignity which emphasize patient autonomy and quality of life rather than the sanctity of human life.

It is widely believed that traditional Judaism believes in the sanctity of human life doctrine. Writing in a recent issue of the *Hastings Center Report*, Ezekiel T. Emanuel said, for example, the following:

> According to this understanding, any intervention that preserves physical existence, regardless of the patient's level of consciousness, mental abilities, or degree of pain, is in the patient's best interests and therefore is necessary. This view is typically associated with the right-to-life movement and the Moral Majority, but it is espoused by many others, ranging from Orthodox Jews and Seventh Day Adventists... ([8], p. 17).

To support his claim, he cited the writings of Rabbi Jacobovits, the Chief Rabbi of England. This is not surprising, since the suggestion that Judaism holds the sanctity of life view was originally advocated by Rabbi Jacobovits, who suggested that Judaism is committed to the view that each moment of human life is of infinite value ([12], p. 276).

This perception of the Judaic position has been accepted within the Jewish community as well. The Federation of Jewish Philanthropies of New York has a Committee on Medical Ethics containing members from all branches of Judaism. It had the following to say about these issues:

> The physician is committed to prolong the life of his patient and to cure him of his illness. Acting in any other capacity, he forfeits his special character and must be judged like any layman who decides to hasten the death of a deformed or critically ill patient. Active euthanasia is an act of homicide running counter to the great philosophical and ethical values which ascribe infinite worth to even residual life. Passive euthanasia... is likewise a failure of the technician to fulfill his oath of office. When the physician can in good conscience declare the patient no longer responsive to his ministration, the physician is then beyond his ability to serve. At that point, and with due consideration of the risk-benefit ratios of possible further intervention,

Baruch A. Brody (ed.), Suicide and Euthanasia, pp. 39–75.
© 1989 *Kluwer Academic Publishers.*

the physician may withdraw from specific therapeutic interventions and leave life in the hands of God ([7], pp. 107–108).

It is not my intention in this essay to argue that this viewpoint is entirely incorrect. Judaism has not traditionally stressed patient autonomy, and it has on the whole opposed suicide and euthanasia. I shall argue, however, that it is not committed to a belief in the sanctity of human life, to a belief that mere physical existence is in the patient's best interest or to a belief that residual life in pain has infinite worth. Such claims totally misrepresent the traditional Jewish position and fail to bring out important aspects of it which are essential to its contemporary elaboration. It also makes it impossible for us to learn from the traditional Judaic discussions an important lesson about the structure of a balanced moral life.

In this essay, I will argue for my position by surveying selected aspects of the rabbinic discussion of suicide and euthanasia. My analysis will explore the origins of, and the reasoning behind, the rabbinic opposition to both, and it will examine some of the exceptions that were recognized over time.

There are aspects of the Judaic discussion of these matters that will not be covered. These include: (1) the extensive rabbinic literature discussing the conditions under which someone who committed suicide should or should not be considered as having been of sound mind, and the implications for burial and mourning practices of such decisions; (2) suicides in Jewish history – including the famous mass suicide at Massada – which are not discussed in rabbinic literature. The former is excluded because it is not relevant to our concern, which is the normative appropriateness of suicide and/or euthanasia, while the latter is excluded because occurrences, however famous, shed no light upon normative attitudes within the mainstream of traditional Jewish thought.

PART I – SUICIDE

A. *The Basic Prohibition*

There are three major rabbinic legal texts from a relatively early period which serve as the basis for all later discussion and which clearly prohibit suicide. We shall examine them one at time. It is worth noting, before we begin our examination, that only one of them refers to the Biblical cases of possible suicide (Samson, Saul and Achitophel). We shall return to this observation later on in our analysis.

The first of these texts is the following Talmudic passage:

There is a disagreement among the Tanaim, for some say that a man is not allowed to hurt himself while others say that he is. Which Tana says that a man is not allowed to hurt himself? Is it the Tana who taught: "But your blood from yourself I will seek punishment (*Genesis* 9:5)"? R. Elazar says, from you yourself I will seek punishment for your blood. Perhaps self-killing is different. Is it the Tana who taught: you may tear garments in mourning over the dead and it is not prohibited as an Emorite custom. R. Elazar said that, I have heard that excessive tearing of garments violates the laws against waste. Certainly that would apply to harming one's body. Perhaps garments are different for that is a loss that cannot be repaired... It is the Tana who taught: "R. Elazar Hakfar said, what do we learn from the verse [about the Nazirite] which says "it will redeem him from the sin that he sinned in himself?" What is his sin? He denied himself wine. We can argue *a fortiori*. If this person who just denied himself wine is considered a sinner, then the person who more fully harmed himself is certainly considered a sinner (*B.K.* 91b).

This text is a very rich text which demands a much fuller analysis than we can offer in this essay. At least the following points must be noted: (1) The text contains a specific prohibition against suicide when R. Elazar derives from part of a series of Biblical verses (*Genesis* 9:5–6) which are prohibitions against killing. R. Elazar's opinion, which is adopted unanimously in the sources, is that suicide is wrong just because it is an illicit form of killing; (2) chapter nine in the book of *Genesis* is the rabbinic source for some of the rules governing the Seven Noahide laws meant to apply equally to all people. As such, the prohibition against suicide is to be viewed as a prohibition applying to all people. I emphasize this point because, in an earlier essay [4], I argued for the need always to be careful, when using Judaic material, to distinguish between material which is part of the Noahide laws and which applies to all, and material which is part of the special convenantal relation between Israel and God. The prohibition against suicide is the former type of material; (3) one of the major arguments for the licitness of suicide is that one's life, like one's body and one's property, is one's own to control, to use, and to dispose of as one sees fit. It is important to note that R. Elazar's prohibition against suicide occurs within the context of a larger passage meant to suggest that Judaism rejects that whole line of thought. One's life, as well as one's body and one's property, is not one's own to use and to dispose of as one sees fit.

The second major rabbinic legal text is also a commentary on the same verse in *Genesis*, but it comes from a later source, *Breishit Rabbah*. Commenting on that verse, the anonymous commentator said:

This [prohibition] includes the person who strangles himself. I might think it applies to the case of Saul. The verse says "but". I might think that it applies to Chananyah, Mishael, and Azaryah. The verse says "but" (*Breishit Rabbah* on 9:5).

There are many points that need to be noted about this text. For our purposes, at least the following are crucial: (1) the text confirms the prohibition against suicide as a form of self-murder; (2) the text makes an important distinction between suicides and those who are willing to allow themselves to be martyred, such as Chananyah, Mishael, and Azaryah. We shall be returning to that distinction and the licitness of martyrdom later on; (3) King Saul's actions are listed as licit, but no account is given as to why they were licit. This will become a matter of great discussion in later literature.

The third of the rabbinic legal texts once more confirms the prohibition against suicide. It is included in the rabbinic collection *Semachot*, whose date is unclear but which certainly contains important material from earlier times. It runs:

If someone commits suicide, we do not perform any rites for him. R. Yishmael says, we say over him "Woe! He has taken his life". R. Akiva says, "Leave him in silence. Neither honor him nor curse him". We do not rend any garments over him, nor take off any shoes, nor eulogize him. But we do line up for the mourners and we do bless them because this honors the living. The rule is: we do whatever honors the living... Who is someone who has killed himself? It is not the person who has gone up to the top of the tree and fallen or the person who has gone up to the top of the roof and fallen. It is the person who says I will go to the top of the roof or the top of the tree and throw myself down and kill myself and we see him do just that. This is the person about whom we presume that he has killed himself (*Semachot* 2:1–2).

This text serves as the starting point for a tremendous literature, which we will not analyze any further, discussing who can clearly be treated as a suicide and what are the implications of that judgment.

In addition to these three legal texts, there are a considerable number of stories told in the Talmud about suicide. But we need to be careful about what use we should make of these stories in deciding about the rabbinic attitude towards suicide. Consider, for example, the tragic story told (*B.B.* 3[b]) about the suicide of the last daughter of the Hasmonean house who did not want to marry Herod, and forget, for now, the question of the historicity of that story. Since the suicide is not the act of a major religious figure, and since there is no commentary in the text about the licitness of the act, we can draw no normative conclusions from the story. As far as I can see, the following stories are usable for drawing some normative conclusions:

a. the refusal of Chanina ben Tradyon to hasten his death (*A.Z.* 18[a]),

b. the mass suicide of the four hundred boys and girls destined for the brothels (*Gittin* 57[b]),
c. the suicide of the mother whose seven sons had been killed (*Gittin* 57[b]),
d. the mass suicide of the young priests (*Ta'anit* 29[a]),
e. the suicide of the Roman officer to save R. Gamliel (*Ta'anit* 29[a]),
f. the suicide of the servant of R. Yehuda haNasi (*Ketubot* 103[b]),
g. the passive suicide of R. Hiyya b. Ashe (*Kiddushin* 81[b]),
h. the passive suicide of R. Elazar b. Durdiya (*A.Z.* 17[a]), and
i. the suicide of the nephew of Yosi b. Yoezer (*Midrash Rabbah* on *Toldot*).

The first three are the most famous, but all of them can be used to raise some normative questions and/or to draw some normative conclusions.

During the Roman persecutions after the destruction of the Second Temple, R. Chanina b. Tradyon continued to teach the law even though such teaching was illicit. He was condemned to be executed. He was wrapped in a Torah Scroll and put on a pile of branches which were set on fire. Cotton soaked in water was placed upon him to slow his dying process. His students suggested that he open his mouth and breathe in the smoke so that he would die quickly. He refused, saying "that it is better that he who gave life should take it and that I should not kill myself". The executioner offered to increase the fire and take off the cotton if R. Chanina promised him eternal life. R. Chanina made that promise. The executioner did it, and then, when R. Chanina died, he jumped into the flame. A *bat kol* came and said that R. Chanina b. Tradyon and his executioner are received into the world to come.

There are many lessons and questions raised by this story. We see once more the theme of the illicitness of suicide, but here it applies even to someone dying in great pain. We see once more the theme of the illicitness of suicide being connected with the idea of God's sovereignty over life. But there are many perplexing questions raised by this story. Why was it permissible for R. Chanina to agree to the hastening of his death by the executioner? Why was it permissible for his executioner to do so? We know that both of their actions were licit because of the judgment announced by the *bat kol*. But why were they licit? We shall need to return to all of these questions.

The second story also occurred during the Roman persecutions. Four hundred boys and girls were taken captive and were being brought in a ship to an unspecified location to be used in brothels. The eldest, in response to a question from the girls, argued from a verse in *Psalms* that they would enter into the world to come even if they threw themselves into the sea. The girls did that. The boys then followed their lead. The Talmud concludes with the observation that it is such cases which the Psalmist had in mind when he said [*Psalms* 44:23] "For you we are killed all the day. We are considered like the sheep being led to the slaughter".

This story, with its accompanying Talmudic comment, becomes a paradigm of a certain type of suicide. But it raises many questions. These young people killed themselves, rather than allowing themselves to be martyred (as did Chananyah, Mishael, and Arzaryah). Is that difference of no moral relevance? And why was it permissible for them to kill themselves but illicit for R. Chanina b. Tradyon to do so? Once more, we shall need to return to these questions.

The Talmudic version of the third story is portrayed as occurring in the times of the Roman persecution. It is clearly based, however, on the story told in *Second Macabees* (Chapter 7) about a mother and seven children martyred during the pre-Hasmonean persecutions. In later texts, the story is restored to its proper time. Our concern is not with the death of the seven children, each of whom allowed himself to be martyred rather than worship an idol. Our concern, rather, is with the death of the mother, for the story ends as follows:

> The mother asked that [the youngest child] be given to her so that she could kiss him. She said to him, "My son, go and say to our father Abraham that he sacrificed on one altar but I sacrificed on seven". She went to a roof and fell down and died. A *bat kol* came and said "The mother rejoices with the sons" (*Gittin* 57[b]).

The story is such a sad and pathetic story that one hesitates even to raise analytic questions about it. Still, the questions are there. The mother was not facing a demand to commit idolatry (so she was not a martyr like her children) nor was she facing a threat to her virtue (so she was not like the 400 boys and girls). Why then was it licit for her to kill herself? Or was her act objectively illicit but forgivable because of her grief? Is that the most we should infer from the words of the *bat kol*?

Similar questions need to be raised about the story of the mass suicide of the young priests during the destruction of the First Temple. The legendary text reads as follows:

> When the First Temple was destroyed, groups of young priests gathered the keys of the Temple in their hands and they went to the roof. They said, God, since we have not been allowed to be faithful treasurers, we return these keys to you. They threw them towards Heaven, and a hand came and caught them. The priests jumped into the fire. Of them, Isaiah asked... (*Taanit* 29[a]).

It is possible, of course, to read the text so that no approval is given to the suicide, even though God himself accepts the keys of the Temple from the young priests. But it is more plausible to see their suicide as acceptable. Yet why is it? The very same questions raised about the death of the mother of the seven sons arise here.

A very different set of questions is raised by the fifth story, the story of the Roman official who kills himself so that R. Gamliel will not be killed. Is suicide permissible to save the life of someone else? Did R. Gamliel sin when he allowed the Roman official to kill himself to save R. Gamliel's life? If not, why not? What is the relation between all of this and the suicide of R. Chanina b. Tradyon's executioner?

A final set of questions is raised by stories (f)–(i), stories that have to do with suicides, active or passive, out of guilt and a desire for penance. The first story deals with an attendant of R. Yehuda HaNasi. On the day of R. Yehuda HaNasi's death a *bat kol* had announced that all present at his death were guaranteed a place in the world to come:

> This attendant used to be there every day. That day, he was not present. When he heard the news, he went to the roof and fell to the ground and died. A *bat kol* came and said that he too had a place in the world to come (*Ketubot* 103[b]).

The second story deals with R. Hiyya b. Ashi, who thought that he had committed fornication with a prostitute (it was actually his wife in disguise). He tried to kill himself. Even after she told him the truth, he said:

> Still I intended a sin. All the rest of his days, he fasted until he died (*Kiddushin* 81[b]).

The third story deals with R. Elazar b. Durdiya, who was notorious as a fornicator. Becoming convinced that he would never be forgiven, he:

> held his head between his knees and he sat and cried until he died. A *bat kol* came and said that he too had a place in the world to come (*A.Z.* 17[a]).

The final story is about the nephew of the early rabbinic leader, Yosi b. Yoezer, who had sinned by riding a horse on the Sabbath. After accepting the admonitions of his uncle, he ingeniously killed himself in a fashion analogous to all four forms of capital punishment recognized in Judaism. His uncle said:

> In a short moment, he preceded me into paradise (*B.R.* on Toldot).

Is penance then a legitimate reason for suicide? Does it make a difference whether it is active or passive suicide? Does it make a difference for what sin one is doing penance? These are all questions unanswered by the texts.

A very interesting phenomenon has emerged from our examination of the Talmudic texts. The legal texts are unanimous in their prohibition of suicide. They view it as a form of murder. They also deny the autonomy principle so often used to permit suicide. Nevertheless, important stories are told which seem to allow for licit suicides in some cases. This seeming paradox will provide the basis for a continuing casuistry in the post-Talmudic period. We turn now to an examination of that casuistry.

B. Suicide and Martyrdom

The obvious place to begin is with the question of suicide and martyrdom. It is clear, from the simple interpretation of the commentary in *Breishit Rabbah* cited above, that the rabbis distinguished killing oneself from acts of martyrdom or of willingness to be martyred as in the biblical case of Chananyah, Mishael, and Azaryah. But three questions require resolution: (a) In what cases is it permissible – or mandatory – to allow oneself to be killed rather than to violate one's religious obligations? (b) May one allow oneself to be killed in other cases or would that be a case of prohibited suicide? (c) May one kill oneself out of the fear that one will commit apostasy under torture and not allow oneself to be martyred? We turn to an examination of these issues.

The classic text about martyrdom is the following:

> R. Yochanan said in the name of R. Shimon b. Yehotzadak: It was decided in the attic in the house of Netazah in Lud that of all the sins in the Torah, if they say to a person do them so as not to be killed, he should do them and not be killed except for idolatry, sexual offenders, and bloodshed... [on idolatry] R. Yochanan follows the opinion of R. Elazar [and not R. Yishmael] who quoted the verse "you shall love the lord your God with all your heart and all your life and all your riches"... About sexual offenses and homicide, he follows the opinion of Reby. As we learned, Reby said... We compare the case of the maiden [who will be violated] to the case of the murderer. Much as in the case of murder, we say be killed and not kill, so, in the case of the maiden, we say be killed and sin not. And how do we know that about murder? It is a matter of logic... Who tells you that your blood is redder than his, maybe his is redder than yours. R. Dimi came and said in the name of R. Yochanan that this rule [of the three exceptions] applies only when there is no general persecution, but in the time of a general persecution, you should be killed rather than violate even a small commandment (*Sanhedrin* 74[a]).

This general opinion of R. Yochanan (a scholar of the third century C.E., but drawing here on earlier material) became the normative rule in Jewish life.

And it is just this framework that provides the basis for our other questions (b) and (c) about the relation between martyrdom and suicide.

As might be excepted, there is considerable controversy about both of these questions. Let us examine each of them separately, beginning with (b). The clearest opponent of martyrdom beyond what is demanded by the law is Maimonides who wrote:

In all cases in which it says you should violate the law and not allow yourself to be killed, the person who allows himself to be killed and does not violate the law is deserving of losing his life (Maimonides, *Yesodei Torah* 5:4).

Presumably, the sin of such a person is that he commits self-murder; otherwise, why should his sin be condemned in such strong terms. It is unclear, however, how serious Maimonides was in putting forward this theory of passive suicide. In any case, Maimonides stood in opposition to the Germanic opinion widespread in the commentaries of Tosafot on the Talmud. A classic text of that opposing approach is the following:

And if a person wishes to be more stringent and risk his life for other commandments he may. This is like the case of R. Abba b. Zimra in the Palestinian Talmud. He was visiting an idolator who said to him that he must eat prohibited food or he would be killed. R. Abba [refused, saying,] "if you want to kill me, kill me". He was being more stringent, because this was probably in private. [Martyrdom would have been mandatory if it had been in public, in accordance with the framework laid down by R. Yochanan.] (*A.Z.* 27^{b} Tos. "Yachol").

A major 16th-century figure, Joseph Karo (the author of the main code of Jewish Law, the *Shulchan Aruch*) offered a clear and precise account of the conflict among these earlier (12th-century) figures in his commentary on the above-cited passage from Maimonides. Karo said:

Our teacher [Maimonides] believes that when the Talmud said "violate and do not be killed", it means that you must violate so that you will not be killed. But many scholars believe that if you allow yourself to be killed, it is a righteous deed. They explain the Talmudic dictum to mean that one is permitted to violate the law so that one will not be killed (*Kesef Mishna* on Maimonides, *Yesodei Torah* 5:4).

This account is an excellent exegetical account of the conflict. Behind the exegetical conflict lies, however, a deeper issue, viz., the conflict between the value of life and the avoidance of suicide, and the value of honoring God by being faithful to his commandments. Not surprisingly, compromise positions emerged, and they too are summarized by Karo in his commentary on another legal code, the Tur. Among the opinions he cites (in his commentary *Bais Yosef* to *Tur Y.D.* 157) are the following:

1. If the order to violate the laws is not done as an act of persecution, but is only done for the benefit of the persecutor, everyone agrees that it would be wrong to allow oneself to be martyred (the opinion of R. Yeruchem);
2. A leading religious leader who sees that the community has become lax in confronting persecutions and wants to set an example may, according to everyone, martyr himself to avoid violating even a small commandment (the opinion of the Nimukei Yosef).

Still further issues and compromises arise when one considers the question of martyrdom in order to fulfill positive commandments, as opposed to the standard case of martyrdom in order to avoid violating negative commandments, but little would be gained for our purposes by following those considerations any further.

We turn then to the even more controversial question (c), the question of committing suicide out of the fear that one will not withstand torture and will violate the law rather than accept martyrdom. Here the classic texts are commentaries on some of the Talmudic stories discussed above. We begin with a commentary of R. Tam (12th Century) on the story of Chanina b. Tradyon. R. Tam was concerned with explaining why he did not hasten his death, while the 400 boys and girls did commit suicide. He offered the following explanation:

> In cases where one is afraid that the idolators will force one to sin by torture that one cannot stand, then it is a mitzva to kill oneself, as in the case of the young people who threw themselves into the sea (*A.Z.* 18ª Tos "V'Al").

R. Chanina b. Tradyon was no longer being offered the choice between life and sinning. He was condemned to death, was being killed, and did not have to fear that he would apostasize to save his life. He had therefore no justification for hastening his death. The four hundred young people were different. They still had to confront the choice between sin and death. They had to fear that they might sin under torture. Their decision to commit suicide was justified.

It is worth noting the language employed by R. Tam. He did not say that it is merely permissible to commit suicide in such a case. Nor did he say that one is obliged to do so. He said that it is the performance of a good deed ("a mitzva").

His opinion was not merely a theoretical opinion. The question arose for the Jewish communities of the Rhineland during the First and Second

Crusades, when the Crusaders, before leaving on their mission, decided to cleanse their homelands of non-believers, and turned on the Jewish communities. The numbers of martyrs were great during that period of time. Some of the most tragic cases involved those who killed both themselves and their children to avoid the possibility of apostasy under torture. This extension of the opinion of R. Tam is found in a commentary from his period on *Genesis* 9:5:

This verse is a warning against those who would kill themselves. It says in *Breishit Rabba* that I might think this applies to Chananyah, Mishael, and Azayah. The verse says "but". This means that I might think that even people like they who gave themselves to martyrdom could not kill themselves if they were afraid that they could not stand the test. "But" tells me that in times of persecution one can allow oneself to be killed and one can kill oneself. The same with Saul... And it is from here that those who killed the children in the time of persecution bring a proof [to justify their action] (*Da'At Zkeinim*, commentary on *Genesis* 9:5).

Leaving aside the tragic dimension of this commentary, a number of fascinating casuistric and exegetical points emerge: (1) the suicide of Saul is justified on the grounds that he feared that he would apostasize under the tortures of his enemies. (2) Chananyah, Mishael, and Azaryah required no justification for allowing themselves to be killed, since that is not suicide. *Breishit Rabbah* is saying that they would have been justified in killing themselves if they feared apostasy under torture. (3) The permission articulated in *Breishit Rabbah* and in R. Tam's writings extends to parents killing their children.

It might be suggested that the Germanic authorities of the 12th and 13th centuries were very supportive of martyrdom and less worried about the issue of suicide. After all, it was they who argued, *contra* Maimonides, that one could commit martyrdom even when it was not required, and, as we have just seen, it was they who supported actively killing oneself to avoid sinning under torture. While this suggestion has some legitimacy, it can be overdone. There were, after all, other authorities within that same community who opposed killing in advance out of the fear of apostasy. To quote once more the same commentary:

Others prohibit the practice. They explain [the remarks of *Breishit Rabbah*] as follows: I might think that this prohibition applies even to Chananyah and his friends who are being martyred. We are told otherwise by "but". Even they, however, cannot kill themselves. I might think that this prohibition applies to Saul... Saul in fact acted against normative opinion. This is the explanation of R.S. b. Abraham who was called Uchman. There was one rabbi who killed many children in the time of the persecutions... A second rabbi who was with him was very angry and called him a murderer.

The first rabbi paid no attention and said that if I am wrong, I should die a cruel death. That happened. The Gentiles stripped his skin and put dirt between it and his flesh. The persecution ceased. If he had not killed the children, they would have lived (*ibid.*).

As is not uncommon in these difficult and tragic casuistrical debates about martyrdom and suicide, no final resolution appeared. But one final attempt was made in the sixteenth century, by the great Polish Talmudist Solomon Luria, to resolve these issues, and we shall end our analysis of these issues by a careful examination of Luria's argument. The text runs as follows:

The law is that one is prohibited to damage or embarrass oneself for the sake of money. And to kill oneself it is prohibited even if one fears that one will be forced to apostasize. And it is certainly prohibited to kill one's children in a time of persecution ... It seems to me that even if he is captured and he is afraid that he will apostasize under torture he should not kill himself. Rather, he should do his best to accept the tortures... One should let oneself be killed rather than commit these sins, and that is not counted as suicide as the Rosh says... it is not suicide to let oneself be killed. But to kill oneself is prohibited. And we find this explicitly in the case of R. Chanina b. Tradyon ... who did ask others to hasten his death. But if one is afraid that they might torture you to testify against many others so that many Jews would die – as some kings have gotten one Jew to testify falsely against many – then it is permissible to kill oneself. And perhaps that is what Saul had in mind when he fell on his sword. He thought that if he were captured, the Children of Israel would not remain passive but would try to save him and many would die... And it is certainly prohibited in times of persecution to kill one's children to save them from apostasy because one cannot even do that for oneself. And if one is not worthy [of martyrdom], he will repent later. In the meantime, he is compelled and is not worthy of punishment. Many forced apostates do repent later, as do some of their children. One can, however, burn down the house and die, and this is not considered suicide but just allowing oneself to die. It is like R. Chanina b. Tradyon who asked that his death be hastened, but who would not kill himself (*Yam Shel Shlomo*, B.K. 8–58).

Luria's brilliant argument involves the following crucial elements: (1) A clear distinction between suicide (causing one's death) and martyrdom (allowing oneself to die), with the former prohibited and the latter encouraged in the appropriate cases. Of particular interest is his view that producing situations in which one's death results – or is quicker – may be allowing one's death rather than causing it. As has often been noted [13], this type of casuistry requires an analysis of causality, something that is currently lacking. (2) A new explanation of the story of Saul – that he was an altruistic suicide – and of the details of the story of R. Chanina b. Tradyon so as to make those stories compatible with Luria's normative claims. (3) Special advice to those confronting martyrdom. The advice comes to this: Try to resist the tortures and gain the glory of martyrdom. If you cannot, you are not a sinner worthy of punishment, because you were compelled to sin, and you can always return to your faith at a later date.

The one crucial element missing in Luria's account is an attempt to explain the case of the 400 young boys and girls, the case which was, after all, the crucial proof case for those, like R. Tam, who argued for the permissibility of suicide to avoid sinning under compulsion. But we shall not extend this analysis any further at this point; we shall turn instead to other cases of permissible suicide.

C. Suicide and Repentance

In the discussions we have examined until now, most of the authors have focused on the initial prohibitory texts and stories (a)–(c). I now want to turn briefly to a series of discussions that focused on stories (f)–(i), the suicides out of remorse and repentance stories.

The crucial figure in this debate is Jacob Reischer (1670–1733), a noted 18th–century, central European Talmudist. Reischer lived in a period of time in which severe penances were regularly sought by those who had sinned and were regularly imposed by rabbinic authorities. It was this sort of milieu which led to the Hassidic reaction and the foundation of the Hassidic movement.

Reischer was asked (*Responsa Shvut Ya'akov* Vol. II, 111) about a person who had killed himself by drowning himself in a river because he had sinned by committing adultery with a married woman. The question was, of course, about burial and mourning practices, but Reischer chose instead to use it as a basis for considering the general issue of the normative permissiveness of killing oneself as an act of repentance.

Reischer had a number of major texts to use in support of his claim that suicide for the sake of repentance is normatively permissible. One that he appealed to in particular was the story of R. Hiyya b. Ashe, who thought that he had committed an act of fornication with a prostitute. Of special significance was his attempt to kill himself by sitting in an oven, indicating that he felt it permissible to actively kill himself. A second story that he appealed to is the story of the nephew of Yosi b. Yoezer, who very clearly killed himself as an act of repentance. Remember that Yosi (a major early pharasiac figure) clearly indicated acceptance of what his nephew had done.

Both stories lend themselves to alternative analyses. Perhaps R. Hiyya was momentarily distraught. Perhaps the nephew of Yosi did what he did incorrectly, but he repented of his sin and his new sin – of suicide – was done when he was not of clear mind. Reischer himself clearly accepts alternative interpretations of the story of R. Elazar b. Durdaya, but he is not willing to do so for the stories he takes to be normatively significant. Of special interest is Reischer's complicated analysis – based on many earlier amplifications – of story (f), the story of the servant of R. Yehuda haNasi.

Interestingly enough, Reischer tentatively concluded that this permissible suicide should be denied some of the mourning rights of a normal person who had died. His argument was characteristically ingenious. The law is that someone who is executed for his sins does not receive the normal mourning rights, for this is part of his atonement. This suicide, suggested Reischer, should not be treated differently.

Many arguments can be offered against Reischer's position. His proof stories all lend themselves to an alternative analysis. There are powerful arguments against his conclusion. They include the following: (1) Even if repentance is a positive commandment, it can be done without killing oneself, and why should this unnecessary form of repentance justify violating the stringent commandment against suicide. (2) There is a Talmudic controversy [in Ta'anit 11[a]] as to whether or not someone who fasts extensively – presumably to help in repentance – is considered a saint or a sinner. But that surely means that both sides would agree that killing for penance is a sin.

D. Altruistic Suicides

In Luria's analysis quoted above, the story of Saul was reinterpreted so that it became an instance of altruistic suicide, with the clear implication being that someone may kill himself to save the lives of many others. Luria did not elaborate on this suggestion, which is quite surprising. We turn then to an analysis of altruistic suicide. Three questions can be raised: (1) May one kill oneself – commit suicide – to save the lives of others? (2) May one allow oneself to die so that other lives will be saved? (3) May one put oneself into conditions of danger in order to save the lives of others?

This set of questions was seen as a very difficult set of questions precisely because Jewish casuistry had placed great significance on the obligation to save the lives of others. The classic text is as follows:

How do we know that, if someone sees a second person drowning in a river or threatened by a wild animal or by brigands, that he is obligated to save that person? It is written, "Thou shall not stand idly by and see the loss of blood of one's friend"

(*Leviticus* 19:16). Do we not derive the obligation from another argument, which runs as follows: how do we know that if someone is about to lose their life [that we must save it]? It is written, "Return it to him" (*Deuteronomy* 22:2). From the second source I would merely learn that he must personally try to save him. But to bother to spend money to hire others, I might think that one is not obligated. The first source tells us that I am (*Sanhedrin* 73ª).

A number of crucial points should be noted about this text. To begin with, The Good-Samaritan Problem, which has been a matter of such controversy in moral philosophy [10], was never an issue in Jewish casuistry. There is both a positive commandment to act to save the lives of others and a negative commandment that one violates if one fails to save that life. Second, the commandment implies that one is obligated to spend one's resources. Much later casuistry – which we will not analyze here – was devoted to the question of how much of one's resources one was obliged to spend, whether the person who was saved had to pay you back, and whether the community as a whole had to reimburse the saver if the saved person could not. But none of that changes the basic fact of the existence of the obligation. Finally, as the later authorities point out, this set of commandments has great significance. To quote Maimonides:

Even though we do not punish [the person who violates these commandments] because there is no action involved in their violation, these commandments are very serious for anyone who saves a life in Israel has acted as though they have saved the entire world (Maimonides, *Laws of Murder*, I:16).

All of these considerations will count as powerful reasons for acting to save lives even at the cost of one's life (or at the cost of a risk to one's life). On the other side, of course, stands the prohibition against suicide. But it is not just that prohibition which stands there, for it would only prohibit killing oneself. There are other prohibitions reinforcing the second side and arguing against putting oneself in danger. A classic text indicating those prohibitions is the following text:

It happened with a righteous man that he was praying and an officer came and said hello. The righteous man did not respond. The officer waited until he finished his prayers and said to him: Does it not say in your Torah "Protect yourself and your life" (*Deuteronomy* 4:9) and does it not say "Watch your life carefully" (*ibid* 15). Why then did not you answer me [I might have killed you]... (*Brachot* 32ᵇ).

The righteous man ultimately justified his behavior, arguing that he could no more interrupt his prayers than could someone interrupt his petition to an ordinary human king. Our interest is not, however, in his response but in his acceptance of the officer's argument, an acceptance that became normative, that one is prohibited to act so as to put one's life in danger. Note once more the opposition to the argument from autonomy to the licitness of putting oneself at danger.

We have here the basis for a powerful casuistrical question: How can we reconcile the obligation to save lives and the obligation to protect one's own life? The classic text with which all discussions begin is a famous early rabbinic controversy:

> Two people were going in the way. Only one of them had a container of water. If only one drank, he would survive to get to his destination. Ben Peturah said that it is better that both should die rather than that one should see the death of the other. R. Akiba came and taught that it says "your brother should live with you" (*Leviticus* 25:36), your life takes precedence over his life (*B.M.* 62[a]).

There are three crucial observations that need to be made immediately about this text. The first is that it does not involve the question of one person killing another to save his own life. The illicitness of this is taken for granted. All would agree that the person without the water should not kill the person with the water to take it. The question raised here is our question of what are the obligations of the person with the water. Should he save his life, or should he allow himself to die to save others? The second is that we have here a case of one life versus one life. Nothing in the text itself addresses the issue of sacrificing one life to save many, the issue raised by the story of Saul according to Luria's interpretation. We shall return to this point below. Finally, the text is entirely silent about what a third person should do if he has enough water for only one other.

Both the opinion of Ben Peturah and of R. Akiva require considerable elucidation. Why did Ben Peturah argue that both should die? His literal argument is that one should not see the death of the other. But what sort of argument is that? And anyway, why should one assume that they will die simultaneously? Or did Ben Peturah have other reasons? Was he suggesting, perhaps, that they should share the water now in the hope that they might both be saved before they died? Was he suggesting, perhaps, that each moment of life is of such great value that saving both lives temporarily is better than saving one life permanently by losing one life more quickly? We cannot really say. Of greater importance, of course, are the fundamental ambiguities in the authoritative opinion of R. Akiva. R. Akiva clearly derived from the verse in question the view that one's life takes precedence over the lives of others (so long as one is not killing them to save one's life). But what does that precedence come to? The following crucial questions remain: (a)

Was R. Akiva merely permitting one to save one's life by not sharing the water, against Ben Peturah, who prohibited it, while allowing one to follow Ben Peturah's opinion, or was R. Akiva saying that one is required to save one's life first? (b) The case here involves the certain death of both if the water is shared. Would it be different if one was merely *risking* one's life to save others? Would that be permissible, even according to R. Akiva? Would that even be obligatory? (c) Would the whole situation be different if there were many lives at stake? Would it be permissible – and perhaps even obligatory – for a person then to give precedence to the lives of others? Could one even commit suicide if by doing so one saved many? We need to turn to other Talmudic and post-Talmudic texts to find answers to these questions.

While question (a) has surprisingly attracted little direct attention, question (b) has attracted a lot, primarily because Joseph Karo (the author of the *Shulchan Aruch*) seemed to claim in his commentaries both on the Code of Maimonides and on the Tur's code that there is a Talmudic text supporting the view that risking one's life to save the life of another is obligatory and that there are good reasons to support such a view. I quote the fuller version of his view:

The Hagohot Maimonit says... that the Palestinian Talmud concludes that one is obliged to put oneself into a risky situation. It seems to me that the reason is that it [the threat to the life of the person] is certain but the other [the threat to the life of the potential saver] is only possible and that he who saves a life in Israel is as though he has saved a full world (*Beth Yosef* on *Tur Choshen Mishat* 426).

Interestingly enough, Karo did not mention this law in his own code.

Where is the text in the Palestinian Talmud? The source usually mentioned is a series of stories at the end of chapter 8 of *Trumot*. The stories occur immediately after the well-known and much discussed conflict between R. Yochanan and Resh Lakish about handing over someone to be killed (*not* actually killing him) to save the lives of many. It seems to me that the stories are very revealing and are deserving of careful analysis:

The government was pursuing Ula B. Kushav. He fled to Rabbi Yehoshua b. Levi in Lud. The government came and said that if you don't give him to us, we will destroy the city. Rabbi Yehoshua b. Levi went to him to convince him to give himself up. Until then, Elijah used to visit him. He stopped. R. Yehoshua b. Levi fasted until Elijah came and said that I do not reveal myself to informers. R. Yehoshua b. Levi replied that I followed the law [quoted just above]. Elijah responded that it is not the law of the righteous. R. Imi was in captivity in a dangerous place. R. Yonasan said that he should prepare his shrouds. Resh Lakish said that I will go and save him by force. He saved him by appeasing the brigands. R. Imi said let us go to our elder and let him pray over them. He prayed that what was in their heart to do to R. Imi should be done to them (end of chapter 8 of Palestinian Talmud, *Trumot*).

Several additional stories are told about dangerous rescue missions performed by other rabbis.

There is little doubt that it is these rescue stories which are the basis of the report that the Palestinian Talmud concluded that one is obliged to put oneself at risk to save the life of others. But the stories, of course, fail to establish such a strong claim. They show at most that one is permitted to do so. Consider carefully the discussion between R. Yonasan and Resh Lakish. It is clear that R. Yonasan did not see any obligation to risk his life to save the captive's life. Resh Lakish said that he would try. That is not to dispute R. Yonasan's claim; it is just to say that it is permissible to do so. Perhaps he viewed it as an act of righteousness, like Elijah's suggestion to R. Yehoshua b. Levi.

Such an approach, most fully developed by R. Naftali Berlin (a major nineteenth-century author), would also be useful to explain another relevant story text. The text deals with the question of how much attention one should pay to rumors. It runs as follows:

There were men from the Galilee about whom there was a rumor that they had killed someone. They came to R. Tarfon and asked him to hide them. He said: What shall I do? If I don't hide you, they might see you and kill you. If I do, didn't the rabbis say that one is suppose to attend to – even if not accept - rumors of bad deeds. Go hide yourself (*Nidah* 61[a]).

What was wrong with hiding them? Rashi claimed in his commentary that it was because they might be guilty and it would be wrong to save them, but this is implausible. Should they be allowed to die just because of a rumor that they are guilty? The better interpretation (given by *Tosafot* in his commentary) is that he was afraid that he would be punished for hiding a guilty party and that he was entitled to depend on the rumor to that extent. The implication is that if he knew for sure that they were innocent, he might have been obliged – or at least permitted – to hide them. To quote Berlin:

According to the strict law, R. Tarfon was not obliged to put himself in danger even it was certain that the rumors were false. But he was permitted to do so as an act of righteousness. But because of these rumors, since we should attend to them, he should not go beyond the law's requirement (commentary on *Shi'iltot* 129).

The opinions which we have examined so far on question (b), which permit or require one to risk one's life to save the life of another, naturally need to clarify how much of a risk is involved and when we reach the point where the case is like that of the two men in the desert. A classic text which deals with this is a ruling of the Radbaz (an Egyptian author of the 16th century):

> One is not obliged to put oneself at risk to save his money, but to save the life of his friend... one is obliged to do so even when there is a risk as it says in the Palestinian Talmud. But, if the risk is very likely, one is not obliged to risk oneself to save another. Even if it is an even chance, one is not obliged. Who says which blood is redder? But if the risk is modest and it is likely that he will save the other without being endangered, then he who does not do so violates the commandment... *(Responsa of Radbaz*, 1582).

Of particular interest in this connection is another ruling of Radbaz (1052), that one is not obliged to lose an organ to save the life of another, and that if it is sufficiently dangerous, one would be a foolishly righteous person to agree, a ruling that has important implications for contemporary questions of organ donations from live donors.

Not all authorities agree with this whole approach. The clearest opposition comes from the medieval author of the code *Issur V'Heter Ha'Aruch*, who said:

> But if the person who is safe will be endangered with him, he should not put himself in danger, since he is now outside of the danger. [This is true] even though he sees the death of his friend. We learn this from the verse, "And your brother will live with you". We find no difference between putting oneself in danger and definitely dying (59:38).

This authority obviously understood R. Akiva as prohibiting one's giving up one's life to save another and also extended his argument directly to cases of possible danger.

The other major source of opposition is an argument from silence, one offered by the commentaries on Karo's code, the *Sulchan Aruch*. They noted that he did not mention the law about risks at the appropriate place in his code – and that many of the other major code authors did not either – and they concluded that he did not agree with that law. A tremendous literature has developed to explain why not, but we shall not pursue it any further here, particularly since much of it is based upon *Radbaz* 1052, disregarding *Radbaz* 1582.

We have seen so far that the following is true of altruistic losses of life: (1) one is not obliged to allow oneself to die so that another life will be saved, and that may even be impermissible; and (2) one may be obliged to put oneself at risk to save the life of another, and even if not, this is probably permissible, at least in cases where the risk is not too high. However, we have not found any evidence to deal with two other questions, the question of actually killing oneself and the question of saving the lives of many others, the very questions raised by Luria's analysis of the story of Saul. We turn to a different literature, having to deal with war, to gain some insight into those issues, for it is only in this literature that I have been able to find an explicit discussion of these issues.

There is a set of laws about Jewish kings discussed in the Talmud and codified by Maimonides, but these laws have always been treated as theoretical, for there have been no independent Jewish kings during the whole time of the discussion. Even where there were internally self-ruling Jewish communities, they were not independent communities which had foreign policies and went to war. In recent years, however, with the rise of an independent Jewish state, questions about war – which have always been discussed in connection with kings - have re-emerged for discussion.

The basic principles of the Jewish law of war were codified by Maimonides in the following text:

> A king should first fight those battles which are a righteous deed. These are the wars against the seven nations [the original inhabitants of Canaan], the war against Amalek, and wars to aid Israel against an enemy attacking them. Afterwards, he may fight non-obligatory wars... For the first type of war, he does not need the permission of the court; he can go out at anytime, and he can force the nation to go. For a merely permissible war, he cannot take the nation out to war without the permission of the high court of 71 (*Laws of Kings* 5:1–2).

It is these concepts which lie behind the casuistrical discussion. After all, in wars many will inevitably die on both sides. How can one go to war – putting one's life at risk – in any case? In particular, how can a king ever take a nation into a merely permissible war? A number of contemporary Israeli authors – most of whom believe that one cannot risk one's life to save a single person – have discussed these questions and have offered the following arguments:

1. There is no question that a person is obliged to risk his life to save the lives of many, or to protect the community, and that is why one can fight in these wars (*Responsa Hachel Yitzchak*, O.H. 39, quoted in [16] p. 16).
2. R. Akiva's principle, derived from the verse "your brother shall live with you", does not apply in times of war. Otherwise, merely permissible wars would not be possible. These are governed by the special rules governing the relation of a king – or a nation's rulers – with those who are ruled (*Responsa Mishpat Cohen*, 142–144, quoted *ibid.*).

There is a tremendous difference between these two reasons, for one applies only in times of war, while the other applies in all cases of risking one's life to save others.

It is of some interest to quote the view of the contemporary Israeli author R. Eliezer Waldenberg, who was specifically asked whether a soldier may or must put himself at risk to save a wounded comrade. He wrote:

> The law of the Torah is that in a time of war in a battlefield the law that your brother shall live with you and the law that your life comes first, does not apply... [The soldier] needs to erase from his heart thoughts of himself. He must join himself as an organic whole with all of his fellow soldiers in the battlefield in a way that each must devote himself to save others from danger even if that leads him to great risk. He must do this not merely when his friend is certainly at danger, but even when he is in a possible danger (*Responsa Tzitz Eliezer* 13–100).

I end this section by noting that none of these authors quotes Luria's explanation of the action of Saul and addresses its implication, that one may actually commit suicide – and not merely risk one's life or allow oneself to die – to save many others, either in times of war or at other times.

E. Suicide from a Lack of Satisfactory Life Prospects

All of our discussions of exceptions to the prohibition against suicide have involved suicides not involving self-oriented reasons (martyrdom, penance, altruism). This is not surprising, for they offer the best possibilities for an exception to the strong prohibition of suicide as a form of self-murder. In the remainder of this paper, we will be looking at self-oriented suicides, primarily those involving patients dying in pain. Most of that discussion will be found in the second half of this paper, joined to the discussion of active and passive euthanasia. In this final section of the first half of the paper, we shall look at one self-oriented type of exception, a person who commits suicide because he feels that life has nothing more to offer.

The beginning of a serious discussion of that issue is in a volume of responsa published in the eighteenth century entitled *Besamim Rosh* and attributed to the 14th-century classic author Asher b. Yehiel. One of the cases dealt with is that of a poor man who had suffered much and who had said for many years that he saw no point in living, and who then clearly committed suicide. The author of the text insisted that this man had not sinned because he was like King Saul, a man who killed himself out of all his troubles and sufferings:

The prohibition against suicide deals with someone who denies the good [in the world] and hates the world. He is like some philosophers who do this to dispute God, who praises His world and sees it as good... They say not. Some of them even say that this [suicide] is good for the soul, for it takes it out of the evil body which is like the soul's grave... (*Besamim Rosh* 345).

In short, for the author of this text, only the person who kills himself out of a philosophical opposition to the general goodness of embodied existence in God's world is a sinner.

One thing can be said on behalf of this position. It is closest to the literal understanding of the story of King Saul, for the text says that Saul said to his weapon-bearer to kill him, for "maybe these men will come and stick me and mock me" (I Samuel 31:4). There is no explicit reference to the thought – discussed above – that this was a suicide connected with martyrdom or with altruism. Indeed, some of the classical commentators on the text specifically offered that interpretation. Thus, Kimchi, commenting on that verse, wrote:

Saul did not sin when he killed himself... because Saul knew that he would die in the war because Samuel told him so and because he saw himself surrounded by the archers and he could not escape. It was better that he killed himself and not be mocked by his enemies (*Radak on First Samuel* 31:4).

But Kimchi's view is confined to this type of case, where the person knew that he would soon be killed both in agony and shame, and it is in no way put forward as a general permission to commit suicide. Even Radak's opinion is contradicted by the classic text of Chanina b. Tradyon cited above. So the opinion of Besamim Rosh is a radical new approach, and was seen that way in its time.

Who is the author of that text? The first (1793) edition attributed it to the great 14th-century scholar Asher b. Yehiel. It is generally accepted ([9], Vol. 4 p. 663), however, that much of the text is a forgery by its editor, Saul Berlin (1740–1794), a somewhat bizarre campfollower of the German Enlightenment. Asher b. Yehiel's own views were less clear, but he probably followed the analysis of Kimchi. So this opinion represents an 18th-century Enlightenment viewpoint rather than an authentic aspect of traditional Jewish thought.

It would be incorrect to leave things with that remark, however, for the forged text had some lasting impact into the nineteenth century. While some were quick to dismiss it, others were motivated by it to examine some crucial issues surrounding the story of Saul. Perhaps the most important 19th-century figure we need to analyze is Ephraim Zalman Margoliot, who devoted a long essay to an analysis of issues surrounding suicide. The essay is primarily concerned with burial rites for people who committed suicide without saying in advance what they were intending to do. Margoliot offered many arguments why such rites should not be withheld from them. I quote the crucial sections:

Since he did not say first, how do we know that he did it in spite. Perhaps he did it as an act of repentance, and all who commit suicide as an act of repentance have done a permissible act... We also find in *Besamim Rosh*, that was recently printed, that a suicide is someone who despises God's good like the philosophers, but someone who says that my life is a burden on me because of my poverty is not a suicide. It is true that his proof from Saul is no proof, as Nachmanides and the other commentators explain. Saul knew that he was going to die because of the prophecy of Samuel, who told him that he and his sons would die. For a short period of time alone, it [killing oneself] is permitted, so that he would not be mocked. Nevertheless, he may be right... We certainly find in the Talmud many who committed suicide out of anguish. As in the case of the woman with her seven sons... It is implausible to say about her that she was afraid that she would be forced to sin, as Tosafot says about the children who jumped into the sea [*Beth Ephraim* on Y.D. #76].

Margoliot seems to have adopted the following position: (a) Like Reischer, he believed that it is permissible to commit suicide as an act of repentance; (b) Saul's suicide was permissible to avoid a mocking and cruel death because he only had a short time to live anyway; (c) other Talmudic texts, such as the story of the mother of the seven children, suggest either that suicide out of great pain is permissible, or at least, as *Besamim Rosh* suggests, that the person is not judged a sinner.

The opinion of *Besamim Rosh* was resurrected in our times by at least one rabbinic authority, Rabbi Ephraim Oshry, in counseling a man who came to him in the ghetto of Kovno two days (October 27, 1941) before the great slaughter of Jews in that ghetto. The man wanted to commit suicide to avoid having to undergo a terrible death and to avoid having to see the death of his family. Rabbi Oshry invoked the authority of *Besamim Rosh* as one of his reasons for allowing the man to commit suicide, and suggested that Chanina b. Tradyon was acting in great saintliness, beyond the demands of the law. It should be noted, however, that the more modest opinion of Kimchi would have been sufficient to justify Rabbi Oshry's lenient opinion.

F. Summary to Part One

We have come a long way from the initial basic prohibition against suicide. The following general themes have emerged from our survey:

1. Suicide is viewed as morally wrong because it is an act of self-murder. The opposition to suicide is part of the general opposition to the view that human autonomy is *the* fundamental moral value.
2. All authorities distinguish killing oneself from allowing oneself to be martyred. While the former is generally prohibited, the latter is mandated in a limited number of cases.
3. There is considerable controversy about allowing oneself to be martyred in other cases and about killing oneself out of fear that one will not stand up to the torture and be martyred in those cases in which martyrdom is mandated.
4. Similar controversy exists over suicide as an act of penance and over altruistic suicides. The most permissible cases are those involving putting oneself at risk of loss of life to save the lives of many fellow soldiers during war.
5. It may be permissible to kill oneself to avoid otherwise inevitable and imminent death by torture and indignity at the hands of one's enemy.

PART II – EUTHANASIA AND SUICIDE IN THE TERMINALLY ILL PATIENT

The question of euthanasia in the terminally ill patient has attracted less attention in the Judaic literature than the question of suicide. Still, there is sufficient material to enable us to examine approaches to the following major questions: (a) May one actively kill a dying patient who is in great pain (active euthanasia), and may that patient kill himself (suicide)? (b) May one withdraw care from a patient who is dying in great pain, and may one withhold care from such a patient when the medical care in question would keep the patient alive for some additional period of time (passive euthanasia)? (c) May one provide pain relief to a dying, suffering patient when there is a significant chance that this pain relief will hasten the death of the patient? In this second part of this essay, we will examine each of these questions separately.

A. Suicide and Active Euthanasia in the Dying Patient

The basic Talmudic text about causing the death of a dying patient is the following.

A dying patient [*gosses*] is like a living person in all matters... One may not bind his jaws, nor may one close up his orifices, nor may one put a vessel of metal or any cooling object on his navel until he dies, as it is written, "until the silver cord is rendered asunder" (Ecclesiastes 12:6). One may not move him nor place him upon the sand or upon salt until he dies. One may not close the eyes of a dying person. Someone who touches or moves the dying person has killed him. R. Meir says, "[he] is compared to a flickering flame. A man who touches the flickering flame extinguishes it. Similarly, someone who closes the eyes of the dying person is considered as though he had killed" (*Semachot* 1:1–4).

There is a second Talmudic text, however, which might be interpreted as raising some questions about point (2). The text runs as follows:

If ten people beat him [the victim] with ten beatings and he died, whether they did it simultaneously or sequentially, they are not punished. R. Yehuda b. Betayra said, if they did it sequentially, the last is punished because he hastened the death. R. Yochanan said: both [authorities] derived their view from the same verse, (Leviticus 24:17): "If a man hits all the soul of another man". The first opinion interpreted "all the soul" to mean [that he is not punished] until all the soul was present. R. Yehuda b. Betayra interpreted "all the soul" to mean [that he is punished] as long as any part of the soul is present. Ravah added: everyone agrees that he is not punished if he is a trayfa [if one of his vital organs has been fatally injured]. Everyone agrees that he is punished if he is dying from an illness not inflicted by human beings. The controversy is when he is dying from an illness inflicted by human beings (*Sanhedrin* 78[a]).

This is a difficult text, one which has attracted considerable attention. The crucial point to keep in mind is that this is not a controversy about the licitness of the actions in question. Both parties agree, for example, that no one is punished if all ten beat the victim simultaneously, but obviously they all agree that doing that is illicit. So this text, as important as it is for understanding the full rabbinic attitude to the dying patient, is not relevant to the question of the moral licitness of killing a dying individual, even one whose vital organs have been fatally injured.

It is important to understand that this view of the illicitness of killing the dying patient is *not* based on a view that life is always better than death. A very important set of discussions about the licitness of praying for the death of a patient brings out this point. The text which is the point of departure is the following text, telling of the death of R. Yehuda haNasi, the author of the *Mishna*:

When she [his handmaid, noted for her wisdom] saw how often he had to go to the privy [he was suffering from bad diarrhea] and how much he was in pain, she prayed that it should be God's will that the immortal [angels, who wanted his death] should win over the mortal [humans, who did not]. The rabbis would not stop praying for his survival. She took a jar and threw it on the ground. They stopped praying. Reby died [*Ketubot* 104[a]].

Notice several crucial points involved in this text: (a) her judgment was that Reby's pain made his life unbearable and that he would be better off dead; (b) she prayed for that death and stopped others from praying for his survival.

One might claim that this is just a story about the handmaid of Reby, and even if she were well-known for her wisdom, perhaps the law does not follow her views in this case. There is, however, another, more authoritative source which is explicit on this point. It is the commentary of R. Nissim of Gerondi (a famous 14th-century author) to a Talmudic passage talking about the merits of visiting the sick and praying on their behalf. The passage is somewhat obscure, but R. Nissim offered the following non-standard interpretation:

> It seems to me that this is what the passage is saying: there are times that one needs to pray about the sick person that he should die since he is in a great deal of pain because of his illness and he cannot live. As it says in *Ketubot* [referring to the above-cited story]... [R. Nissim on *Nedarim* 40[a]].

Some later authorities, while agreeing with R. Nissim, expressed the view that only disinterested parties may pray for the death of such a patient. Still other authorities disagree entirely (R. Eliezer Waldenberg, *Ramat Rachel* #5).

Even those who disagreed, however, clearly understood the following crucial point: There is a basic prohibition of killing the dying person because it is murder, as we see in the above-cited text from *Semachot*. Praying that a person should die is *not* murder, so it is not prohibited for that reason. The controversy is about whether or not a short period of life in pain is worth living. R. Nissim and his followers thought that it might not be; the others disagreed.

We can conclude that active euthanasia of a dying patient, even one who is in great pain, is prohibited in Jewish law because it is an act of killing, even if the goal in question – the death of the patient – is viewed as desirable. Desirable ends do not justify all means unless one is an act-consequentialist, and there is little doubt that Jewish law is not an act-consequentialist morality. Note, finally, that although no one explicitly extends this prohibition to suicide, the argumentation would seem to apply there as well. It might be suggested that one could apply Kimchi's opinion about Saul to such cases as well, but I have not discovered any author who has done so.

B. Passive Euthanasia (I) – Withdrawing That Which is Keeping the Patient Alive

The standard bioethics literature contains extensive discussions of the moral significance of the distinction between killing a patient, withdrawing care, which is prolonging a patient's life, so that the patient can die, and not initiating care which will keep him alive for some period of time. These differences are not new. We find an extensive discussion of them in late medieval and post-medieval rabbinic discussions, and we turn now to an analysis of those discussions.

One preliminary point needs to be noted. In all of these cases, the intention of the actor is that the patient should die so that he or she will not suffer any more. In all of these cases, the judgment of the actor is that the patient would be better off dead. Moral theories which emphasize intentions [11] or consequences [5] will have difficulty distinguishing these cases. Jewish casuistry has always distinguished them. We can understand this in light of what we saw in the previous section of this paper. The Jewish opposition to active euthanasia is an opposition to killing the patient, i.e., to causing the patient's death. When that deontological constraint is not violated, other possibilities may be licit.

The classic texts which are the points of departure for the discussion of withdrawing that which is preventing the death of a patient are two commentaries of the 16th-century Polish commentator R. Moshe Isserles. One occurs in his commentary on the *Shulchan Aruch*, and one occurs on his commentary on the code of the Tur. In his commentary on the *Shulchan Aruch*, he said:

> It is also prohibited to cause the person to die more quickly. As in the case of a person who has been dying for a long time and cannot [die]. It is prohibited to [cause his earlier death]... but if there is something that is causing a delay in the death, as for example if there is a noise nearby... or some salt on his tongue, it is permitted to take them away, for that is not an act but only the taking away of what prevents the death (Isserles on *Sulchan Aruch* Y.D. 339: 1, quoting the Hagaoth Alfasi).

The argument is even clearer in his commentary on the parallel passage in the Tur's code. There, quoting once more the Hagaoth Alfasi, he wrote:

> It is certainly prohibited to do something that will cause him not to die... All such things, one is permitted to take away. But to do something which causes his death to be quicker is prohibited [*Darkei Moshe* on Tur Y.D. 339:1].

Isserles's position comes to this: It is prohibited to do that which will cause the patient's death to be delayed, presumably because the patient is suffering and will die soon anyway. Therefore, if one has already done that thing, and it is delaying the patient's death, one may take it away. Then, the patient dies from his illness. All of this is different from cases of causing the patient's death to occur more quickly.

Several additional points about this position need to be made. To begin with, it requires a good theory of causality, one which distinguishes cases of causing the death to occur more quickly from cases of merely allowing the death to occur. The supercommentaries on Isserles (especially *Shach* and *Taz*) struggled with this problem without any sense of a clear conclusion emerging. We shall return to it below. Second, it is clear that Isserles was in agreement with the opinion of Nissim of Gerondi, discussed in the last section, that death is sometimes a blessing; otherwise, why would he prohibit doing something that will cause the patient not to die. Third, Isserles did not specify how imminent the death must be of the patient who is dying before one may remove that which is delaying his death. The view has recently been expressed by Rabbi David Bleich ([3], p. 141) and by Dr. Fred Rosner ([14], p. 200) that this opinion only applies to a patient who cannot be maintained – no matter how aggressive the care – longer than 72 hours. The text of Isserles conclusively proves otherwise, because it begins, "as in the case of a man who has been dying *for a long time*". Their proof is based on misunderstandings already clarified in the supercommentary *Pitchei Teshuva*. It is incorrect then to see Isserles's opinion as confined to patients whose death is that imminent no matter what is done.

In the current public discussions of withdrawing care which is keeping a patient alive, the standard example is that of "pulling the plug" on a respirator-dependent dying patient. From the perspective of Isserles's opinion, the question becomes whether one is taking away that which is preventing his death or causing him to die more quickly. This very question was recently put by Professor David Mayer, the director of the Sharei Tzedek Hospital in Jerusalem, to R. Eliezer Waldenberg, the very important contemporary Israeli rabbinic figure. R. Waldenberg's response is lengthy and complicated, but it initially seems to allow removal of a respirator. To quote R. Waldenberg:

The heart of the difference [introduced by Isserles] is between [on the one hand] when his action only takes away the outside cause that brings him no life of his own, where it is prohibited to apply such a thing which prevents his death, and where it is then permissible to take away that object even if it requires an action, and where [on the other hand] he still has some independent life of his own, and when his action causes a prohibited act of hastening the patient's death... According to Isserles it is permitted to act so as to take away the respirator only when it interrupts the living functions that come from the outside and it doesn't do anything to stop independent living (*Tzitz Eliezer* Vol. 13 #89).

In fact, however, he is doing just the opposite. What he is actually claiming is that Isserles's permission to withdraw care keeping the patient alive is only when the patient shows no ability to perform life-functions on his own. If he does, withdrawing care such as a respirator is hastening his death.

I believe, however, that this is an implausible interpretation of Isserles's opinion. After all, Isserles was permitting the withdrawal of care when it is prohibited to provide it, presumably because the patient has been living in great pain for a period of time. If so, the patient (in this time) must have still been capable of undergoing some independent life-functions. To be sure, it is very hard – perhaps impossible – to distinguish between causing the hastening of the patient's death and removing that which is causing a prolonging of his dying process. It is this difficulty, I submit, which led R. Waldenberg to his implausible interpretation. Perhaps, at the end, it would have been better if Isserles had distinguished between causing the death of the patient (which is prohibited, even if he will die shortly anyway) and merely allowing the disease process to cause his death by removing impediments (even if that hastens the death). The problem arises only because Isserles, following the earlier texts, prohibits, "causing the person to die more quickly", which seems to suggest that the action is prohibited when it causes the hastening of the death, despite the fact that the underlying cause of the death is the disease process. Even the best theory of causality may not be able to make out the distinction between removing impediments and causing the *hastening* of death.

C. Passive Euthanasia (II) – Providing Further Care to Save the Life of the Patient

Section B began our examination of the question of passive euthanasia. In this section, we continue that examination, looking now at the question of whether one is obliged to do everything possible to keep a patient alive, even if the patient is dying and in a great deal of pain.

Here, as in many other cases, there are arguments on both sides. On the one hand, there is the obligation to preserve life, an obligation which has, as we saw in Second D of Part I, a great deal of force in Jewish law. On the other hand, there is the thought, expressed in the above-cited remark of Isserles, that it would be wrong to cause a prolongation in the dying process of a patient in pain. We will need to examine both of these arguments carefully.

Is there any obligation to save the life of a dying person? Certain Talmudic texts suggest that there is such an obligation. The most crucial one occurs in a discussion of violating the Sabbath to save someone's life. The Mishna (*Yoma* 8:7) says that one can clear away on the Sabbath a building that has fallen on someone. It goes on to say: "If they find him alive, they clear [the rest of the debris] from him". The Talmud commented on this passage as follows:

> If they find him alive, it is obvious [that they should clear away the rest]. We need this text to tell us [that we clear it away] even to save his life for a very short time [*Yoma* 85[a]].

A somewhat similar concept is expressed in the commentary of Tosafot elsewhere. The commentary reads as follows:

> Someone who kills a dying person whose death has been caused by human intervention is not punished as it says... most dying patients die. Still, we violate the Sabbath for such a dying person as it says in Yoma that we don't act according to the majority in cases when we are dealing with saving a life (*Niddah* 44[b] Tos. "Ihu").

Although these two texts express similar thoughts, the reasoning process is different. The Talmudic argument is that saving a life for a very short period of time justifies violating the Sabbath, presumably because of the great significance of saving even the life of someone who is dying. Tosafot's argument refers instead to the possibility that the patient, against our expectations, may survive, and that this possibility justifies violating the Sabbath. It is unclear why Tosafot felt the need to add this argument.

The argument which emerges from these passages is straightforward: if a short period of additional life is sufficiently important to justify violating the Sabbath, is that not because it is still included in the commandment to save life? And if that commandment applies to a dying person with a short period of life to live, then do we not have an obligation to provide the medical care in question? Just that argument was offered by R. Eliezer Waldenberg:

> We learn from these texts [about violating the Sabbath] to our case that there is certainly an obligation to do what one can on a weekday to save the life of the dying patient with any medication possible, even for a short time only. The moments of life of a man in this world are very precious for some obtain the world to come in one moment with a thought of repentance... and we are certainly obligated to use any

means to save the person because of the slight possibility of truly saving his life as a small number do live (*Ramat Rachel* #28).

Note parenthetically that R. Waldenberg was using both of the arguments mentioned above, the Talmudic argument and the argument used by Tosafot. Note also that this text provides an explanation of R. Waldenberg's already cited opinions opposing praying for the death of the patient and opposing withdrawing care from the dying patient until he can no longer perform life functions on his own.

So much for the argument on the one side. We turn now to the argument on the other side, the argument that it would be wrong to save the life of the person for a short time if that means that he lives in further pain. That argument is, of course, supported by the remarks of Isserles about it being permissible to withdraw that which is prolonging the dying process and which it was prohibited to apply to the patient initially. Just this argument was offered by R. Moshe Feinstein, the recently deceased American author:

It seems to me that since there is in this medical care only the capacity to extend his life a short time, if this short life that he will live with this medical help is with a lot of pain, it is prohibited [to use this medical care], and even the *Shvut Ya'akov* [who in general thought that there was an obligation to save life for a short time] would agree. Probably this is the reason that it is permissible to take away that which prevents the death... Certainly it is prohibited to use means to lengthen life for a short time if it will be in pain (*Igrot Moshe* on Y.D. II, 174).

Presumably, those who adopt this position will differentiate this case from the cases mentioned in the Talmud and in Tosafot by claiming that those cases did not involve the person's dying in pain.

There are a number of aspects of this discussion that require further elucidation by future authors: (1) What is the relevance of the views of the patient himself? Would R. Feinstein consider it prohibited to continue to keep the patient alive – in great pain – if the patient requests that this be done, because he has unfinished agendas or because he values his life even in pain? Would R. Waldenberg consider it obligatory to keep the patient alive – in great pain – if the patient asked that further medical care be withheld? Neither author, in this context, addresses the question of the significance of the patient's view. (2) Is there any distinction between forms of care? Would R. Feinstein allow the patient not to be fed? Does R. Waldenberg insist on cardiopulmonary resuscitation in every case? In short, many of the familiar issues in the bioethics literature about withholding care needed further elucidation by the adherents of these two positions.

One very recent text has addressed these questions. R. Shlomon Zev

Auerbach, a newly-emerging major figure in Jerusalem, wrote about them in 1979 as follows:

> Some believe that the same way we violate the Sabbath to save life for a short time, so we are obligated to force the sick person [to be treated], because he doesn't own himself so that he can waive even a moment of life. Probably, however, if the sick person is suffering a lot of physical pain, or even if he is in great psychological pain, I believe that we are obliged to give him food and oxygen for breathing against his wishes, but we can withhold treatment that causes pain if he requests it. But if he is pious and will not become despondent, it is desirable to explain to him that one moment of repentance in this world is more valuable than all of the world to come ([2], p. 131).

No argumentation is presented, so we have to reconstruct it. I think that the following seems to be his view: (a) R. Waldenberg's view is essentially correct, and even if he is in great pain, there certainly is no prohibition to keeping him alive. (b) Still, the patient can refuse the offered care, and the shortness of his remaining life and his pain exempt him from the rule that people are not permitted to refuse life-saving care, a rule normally grounded in the downplaying of autonomy in Jewish thought. (c) That permission applies primarily to care which is painful and does not apply to supplying basic life needs such as oxygen and food. Naturally, there is much more that needs to be said about these issues, and about the merits of R. Auerbach's attempt to compromise between the positions of R. Feinstein and R. Waldenberg, but his opinion indicates a line of approach which may become popular in traditional circles in the years to come.

D. Aggressive Pain Relief and the Borderline Between Active and Passive Euthanasia

We turn finally to the last of the questions we need to address, the question of whether or not it is permissible to aggressively manage the pain of a dying patient, even if that aggressive pain relief may result in the patient's dying sooner. The most common example of this problem is, of course, the use of pain relievers such as morphine in dying patients in respiratory distress, where the morphine may exacerbate the respiratory difficulties.

Several arguments might be offered against the licitness of such pain relief. They include the following: (a) This type of pain relief might cause the death of the patient, and we have seen that active euthanasia is clearly prohibited in Jewish law. (b) Even if this is not seen as a cause of active euthanasia, how can it be permitted if it means that the patient's life is shortened? Have we not seen that saving a patient's life for a short period of time is of great value?

In order to analyze these issues, we need to begin by analyzing an important Talmudic text about risking short remaining periods of life. It occurs in the context of a rabbinic prohibition against using non-certified pagan physicians who were suspected of causing the death of their patients. The text runs as follows:

Rava said in the name of R. Yochanan and some say that R. Hisda said in the name of R. Yochanan: If he will certainly die [without that care], he can seek a cure from him. Is there not [a loss] of a short period of life? We are not concerned about that loss (*A.Z.* 27^{b}).

The rabbis prove that claim by reference to the story in II Kings 7, of the four lepers who risked the short period of life they had left without food to enter a city in which they might be killed in the hope of finding food that would keep them alive. There is, despite this proof, an obvious difficulty with this text. We saw in the previous section that the Talmud specifically allows the violation of the Sabbath to save a person's life for a short period of time. Why then does this text say that we do not worry about short periods of remaining life? Tosafot asked that question and offered an analysis which became the basis for all discussions of our issues:

We can say that in both cases we do what is good for him [the dying person]. There, if we don't worry [about saving his life by moving the building off him] he will die. Here, if we do worry and don't allow him to be cured by the pagan, he will certainly die. In both cases, we go away from the certain to follow what offers a possibility (Tos. *ad. loc.* "L'chaye").

The argument is straightforward: we save his remaining life rather than letting him certainly die, even if that means violating the Sabbath, but we risk his remaining life to possibly secure the benefit of saving his life for a long time. These two policies are compatible with each other.

The principle which clearly emerges from this text is that one can risk short remaining periods of life for the patient's benefit, but this text refers to the benefit of possibly curing the patient so that his life is lengthened considerably. Naturally, this does not resolve the issue of whether one can risk the patient's short remaining period of life for other benefits (e.g., pain relief). We will return to that issue in a moment. First, we need to explain the implications of this text, whose validity has been widely accepted ([1], pp. 44–48) with particular reference to the significance of the wishes of the patient.

The question can be put very simply: does the text imply that it is permissible to risk the remaining life of the dying patient to try to save his life if the patient wishes to take that risk, or does it imply that it is mandatory to take that risk? That question has recently been raised by a number of authors, and one who has responded to it is R. Shlomo Zev Auerbach of Jerusalem. Dr. Abraham (of Shaarei Tzekek Hospital) tells of a fifty-year-old patient who had already lost one gangrenous leg due to problems with circulation in his extremities and who was suffering great pain in his other leg. The patient would die in the near future without a second amputation, but he might die on the table, and the surgery would not help his underlying problems. The patient was opposed to surgery, in part because of the prospect of post-operative pains, but in large part because he did not want to live that way. Dr. Abraham goes on to say:

> I asked R. Auerbach what the opinion of the Torah is in such a case. He ruled that one should certainly not operate against the wishes of the patient nor should one try to convince him to agree to the surgery since we are dealing with a dangerous surgery which will only increase his pain without any hope of curing him permanently (*op. cit.* p. 48).

Several points emerge clearly from this text: (a) R. Auerbach's opinion is that it is permissible, but not obligatory, to risk one's remaining life in order to extend life, and that, at least in the case where there is much pain and no hope of a real cure, a patient can refuse to undergo the medical treatment in question. (b) It is unclear why this case is different from the one cited above, where R. Auerbach thought it appropriate to urge the patient to allow further therapy to prolong his life, even though he felt that the patient was permitted to refuse it. (c) It remains an open question as to whether the patient may refuse the care when it might really cure his disease and prolong his life indefinitely.

We turn more briefly to a second preliminary question which we need to examine, viz., may a patient undergo risking surgery to avoid pain or discomfort (physical or psychological) when the patient is not dying and is risking his indefinite life? This question was first raised in the eighteenth century by R. Jacob Emden, who offered no definitive answer, although he seemed inclined to allow the patient to undergo the surgery if he wished it, treating it as a sufficient benefit. A more positive approach was offered in the nineteenth century in Responsa *Avnei Nezer*. An even more positive appraisal of this is found in twentieth-century authors, many (but not all) of whom approve plastic surgery to deal with cosmetic problems ([6], p. 225; [1], p. 48).

Having examined these preliminary questions, we turn now to our issue. The major author who addressed this question is R. Eliezer Waldenberg. He offered these arguments for the permissibility of aggressive pain relief: (a) If it is permissible to risk indefinite life to save the patient from pain, as it may well be, it is surely permitted to risk his short remaining period of life to save him from pain. (2) Pain can cause a shortening of the patient's life, much as morphine can, so we can not tell which course of action preserves life, but the morphine at least avoids pain. (3) All medical care involves some risk of life, and the permissibility of medical care is the permissibility of risking life to attain legitimate goals:

> Nachmanides, in his *Torat Ha'Adam*, explains that the Tora had to give permission [for doctors to cure] because there is no medical care without a risk that what cures one will kill another. Therefore, giving morphine, even if it only quiets pain, is part of what is included in medicine... and it is permitted to give it even though it may hasten his death (*Tzit Eliezer* Vol. 13 #87).

There are several crucial points to be noted here: (a) This is an argument being offered by the same authority who requires all medical care which can prolong the life of a person and who forbids withdrawing care until the patient has no capacity for independent life. (b) The argument rests on risks and benefits, drawing on the idea that risking the patient's remaining life may be justified by the benefit of relief of pain. Even on this account, human life is not the only value. (c) Although R. Waldenberg does mention the fact that the physician is not intending the death of the patient, it is no part of his argument for the permissibility of pain-relief. There is no appeal here to the Catholic doctrine of double effect [11].

E. Summary of Part II

The following themes have emerged from our survey of the discussions of euthanasia and of suicide by terminally ill patients:

1. Suicide and active euthanasia are prohibited because they are illicit acts of killing, but *not* because each moment of human life is definitely judged to be worthwhile. Many authorities concede that the death of some patients is a blessing.

2. As a result, it may well be prohibited to provide care which prolongs the suffering of a dying patient, and permissible to withdraw that care if it has been provided, although some authorities would limit that permission to a very few cases.
3. Because one may risk short remaining periods of life for the patient's benefit, one may provide active pain relief even if this risks the patient's dying sooner.
4. There is a suggestion among the most recent authors that a patient's wishes play a significant role in some of these decisions.

PART III

We have examined a large number of issues as discussed by many leading authorities. I now want to return to the question raised at the beginning of this essay, viz., whether traditional Judaism is committed to the doctrine of the sanctity of human life. I think that it is evident from what we have seen that it is not. No doctrine of the sanctity of human life could justify penitential acts of suicide. No doctrine of the sanctity of human life could justify killing oneself to avoid sinning under coercion. No doctrine of the sanctity of human life could justify killing oneself to avoid a mocking and cruel death at the hands of one's enemy. No doctrine of the sanctity of human life could justify withholding care that would keep a dying person alive but in pain. No doctrine of the sanctity of human life could justify risking a loss of life to avoid pain. Major traditional Jewish authorities have justified all of the above. None of them could have been committed to a belief in the sanctity of human life.

What then is the traditional Judaic opinion? I think that it is comprised of several major elements: (a) a belief in the great value of preserving human life and in the corresponding obligation to come to the aid of those whose life is threatened; (b) a belief in a nearly absolute prohibition against taking the life of the innocent, one's own life or the life of others, even if that person is dying anyway; (c) a belief in a variety of other values (such as integrity in one's commitment to God's law and the legitimacy of avoiding cruel and painful deaths) which may sometimes take precedence over the value of continued living, or even over the prohibition of not killing. The rabbinic figures we have been studying, like all great casuistrists, were trying to balance these many values. They often disagreed among themselves about how these values should be balanced. They all understood, however, that there were many values to be balanced.

The contemporary debate about death and dying has degenerated into a conflict between two simple-minded positions, one which emphasizes a monolithic belief in the sanctity of individual choice and one which emphasizes a monolithic belief in the sanctity of continued biologic existence. Rabbinic casuistry wisely avoided both. Different people may disagree about which values are in conflict and about how they ought to be balanced. They may agree or disagree with particular rabbinic balancings. They all need to learn from the rabbinic discussions, as I have tried in a recent book [5], the merits of avoiding monolithic simple-minded positions of all sorts.

BIBLIOGRAPHY

[1] Abraham, A.S.: 1985, *Nishmat Avraham on Y.D.*, privately printed, Jerusalem.
[2] Averbach, S.Z.: 1981, 'Caring for A Dying Patient', in M. Hershler (ed.), *Halacha U'Refuah*, Vol II, Regensburg Institute, Jerusalem.
[3] Bleich, J.D.: 1981, *Judaism and Healing*, Ktav Publishing, Hoboken.
[4] Brody, B.: 1983, 'The Use of Halakhic Material in Discussions of Medical Ethics', *The Journal of Medicine and Philosophy* 8, 317–328.
[5] Brody, B.: 1988, *Life and Death Decision-Making*, Oxford University Press, New York.
[6] Brown, S.: 1978, *Shearim Metzuyanim B'Halacha*, Vol. 4, Feldheim, New York.
[7] Committee on Medical Ethics: *Compendium on Medical Ethics*, 6th ed., Federation of Jewish Philanthropies, New York.
[8] Emanuel, Ezekiel T.: 1987, 'A Communal Vision of Care for Incompetent Patients', *Hastings Center Report* 17, 15–20.
[9] *Encyclopedia Judaica*: 1973, Keter Publishing House, Jerusalem.
[10] Feinberg, J.: 1984, *Harm to Others*, Oxford University Press, New York, Chapter 4.
[11] Grisez, G. and Boyle, J.: 1979, *Life and Death With Liberty and Justice*, University of Notre Dame Press, Notre Dame.
[12] Jacobovits, I: 1975, *Jewish Medical Ethics*, 2nd edition, Bloch, New York.
[13] Mack, E., 1989, 'Moral Rights and Causal Casuistry', in B. Brody (ed.), *Moral Theory and Moral Judgments in Medical Ethics*, Kluwer Academic Publishers, Dordrecht.
[14] Rosner, F.: 1986, *Modern Medicine and Jewish Ethics*, Ktav Publishing, Hoboken.
[15] Singer, P.: 1979, *Practical Ethics*, Cambridge University Press, New York.
[16] Zeven, S.Y.: 1957, *Le'Or HaHalacha*, 2nd ed., Zioni Publishing, Tel Aviv.

Baylor College of Medicine,
Houston, Texas, U.S.A.

DARREL W. AMUNDSEN

SUICIDE AND EARLY CHRISTIAN VALUES

The Nature of the Problem and the Scope of this Essay

The development of an early Christian position on suicide presents some interesting problems for the historian. In two respects it resembles the ethics of abortion: *First*, Scripture is silent about both.[1] *Second*, arguments against the moral permissibility of either, formulated inferentially from Scripture by the church fathers, are easily rejected as being heuristic. Suicide, however, differs from abortion in that while even the earliest non-canonical Christian literature denounces abortion in unequivocal terms, condemnations of suicide by the early church fathers are relatively rare and hardly unequivocal.

There are at least three reasons for the comparative rarity and the equivocal nature of these condemnations. *First*, such condemnations of suicide by the church fathers as are extant were not part of the broad moral condemnation leveled by early Christian authors against what they regarded as the depravity of pagan society. The moral indignation of the early Christian community, particularly as it was directed against abortion and infanticide, received much of its vigor from the perceived helplessness of the victim, whether a fetus or an infant. Even the occasional condemnation of contraception was motivated in part by concern for the victim (in this case potential life). So also with gladiatorial combat and extremely cruel executions, in which the pagans needed victims to satisfy their lust for blood. Even acts of sexual immorality were often seen as involving victims, for the greatest indignation of Christian authors was reserved for the forced prostitution of both female and male slaves, helpless victims of pagan depravity, and even outside the brothel sexual immorality generally involved more than one individual in the act, hence the potential for there being unwilling victims of others' depravity. It is especially the helplessness of the victims of others' sins that increased the extent of moral indignation to the level so frequently encountered in Christian literature. Suicide did not arouse the same kinds of passionate denunciation, for the act was not seen as one in which an innocent party was victimized by another, but rather as an act in which one harms only oneself.

Second, the ambiguities that arise in modern (and, to an extent, ancient) discussions of suicide tend to obfuscate the issues. The ethics of suicide in early Christianity is more ambiguous than various other ethical issues such as

Baruch A. Brody (ed.), Suicide and Euthanasia, pp. 77–153.
© 1989 *Kluwer Academic Publishers.*

infanticide or abortion. Condemnations of infanticide and abortion by early Christians were unequivocal: no exceptions were even discussed. Furthermore, there was no ambiguity regarding what constituted infanticide or what constituted abortion.[2] But as already mentioned, condemnations of suicide were comparatively rare and were hardly unequivocal. One kind of suicide was approved, at least by some sources: virgins (and even married women) facing sexual assault were lauded by some church fathers for taking their own lives to avoid defilement. Such acts can only be regarded as suicide unless seen through the much later grid of double effect. But some other conditions are exceedingly ambiguous. Is severe asceticism that incidentally but not intentionally results in death suicide? And what of martyrdom?

Discussions of what constitutes suicide flourish today. Such discussions often show a lack of precision in defining the English word 'suicide' and in delineating and circumscribing the concept usually conveyed by that word. The situation in the ancient world permitted even more confusion since neither Greek nor Latin had a specific word for suicide.[3] Because of the definitional and conceptual ambiguities that have helped to foster the current debate on suicide and that have perhaps been rendered even more muddled by it, one may anachronistically read into ancient sources ambiguities quite alien to the latter's concerns. If one regards as suicide any failure to exploit every conceivable expedient to preserve one's life, one subsumes an exceedingly broad range of motivations, priorities, ideals, decisions, and actions under a rubric that then becomes nearly meaningless.

Persecution in its most violent or extreme forms may result in the death of the one persecuted. Let us consider three distinct categories of people thus persecuted. *First*, one may be persecuted, even put to death, for a position that he does not hold or of which he has recanted or would recant if given the opportunity. *Second*, one may also be put to death for a stance of which he could recant but refuses to do so. Willingness to die for a cause is not typically viewed as suicide; otherwise all martyrs, even all soldiers who die in the line of duty, are suicides. *Third*, one who provokes persecution with the intent to be martyred or actively seeks or volunteers for martyrdom may well be labelled a suicide. The first category is irrelevant to our study. The second is certainly germane. Indeed, the very imperative to bear witness to the gospel and the consequent liability to persecution for Christ's followers significantly increased their chances of being martyred when Christianity was an illicit religion. Drawing the line between the second category and the third presents some conceptual difficulties.

One obscures the past rather than clarifies it when with one broad and undiscriminating stroke one labels as suicide *all* those typically called martyrs. A recent study displays a lack of historical discrimination by grouping into one category Donatist Circumcellions of the fourth and fifth centuries, who persistently attempted to provoke Catholic authorities to put them to death; Christians who, before the legalization of Christianity, refused to blaspheme Christ in order to escape execution; "the martyrs who *permitted* themselves to be devoured by starving beasts in Nero's arena" (my emphasis, [30], p. 145). The latter, who were scapegoats put to death after the great fire that destroyed much of Rome in A.D. 64, neither provoked the authorities to execute them nor were likely allowed the opportunity to recant. Much greater circumspection than this must be used in any meaningful historical study. Furthermore, one manifests intellectual arrogance of the worst kind when one analyzes a broad range of martyrs and, employing the ephemeral jargon of current psychological models, assures his readers that these martyrs were motivated by self-punitive, aggressive, erotic, masochistic, narcissistic, and exhibitionistic drives ([30], pp. 146–153). Such mysteries I shall not seek to penetrate.

The various ambiguities mentioned above combined with some serious misunderstanding of some basic tenets of Christianity and an ignorance of patristics, not to mention the New Testament, have led some modern scholars to highly distorted conclusions about early Christian attitudes toward suicide. The following are typical:

There is no condemnation of suicide in the New Testament, and little to be found among the early Christians, who were, indeed morbidly obsessed with death.... The Christian belief was that life on earth was important only as a preparation for the hereafter; the supreme duty was to avoid sin, which would result in perpetual punishment. Since all natural desires tended toward sin, the risk of failure was great. Many Christians, therefore, committed suicide for fear of falling before temptation. It was especially good if the believer could commit suicide by provoking infidels to martyr him, or by austerities so severe that they undermined the constitution, but in the last resort he might do away with himself directly ([43], pp. 254–255).

Even the most stoical Romans committed suicide only as a last resort; they at least waited until their lives had become intolerable. But in the primitive Church, life was intolerable whatever its conditions. Why, then, live unredeemed when heavenly bliss is only a knife stroke away? Christian teaching was at first a powerful incitement to suicide ([1], p. 25).

Christianity *invites* suicide in a way in which other major religions do not... ([8], p. 29). The lure exerted by the promise of reunion with the deceased, release of the soul, the rewards of martyrdom, and the attainment of the highest spiritual states, including union with God, all occur in Christianity.... Thus the question of the permissibility of suicide arises, though often only inchoately, for any sincere believer in a religious tradition of this sort, whether that individual's present life is a happy one or filled with suffering. Religious suicide is not always a matter of despair; it is often a matter of zeal. The general problem presented by the promise of a better afterlife may be strongest in Christianity, since the afterlife of spiritual bliss depicted by Christianity is a particularly powerful attraction ([8], pp. 72–73).

Augustine is usually credited with being the architect of the Christian condemnation of suicide. For example, "... the early Christian community appeared to be on the verge of complete self-decimation in voluntary martyrdom and suicide until Augustine took a firm position against such practices" ([8], p. 89). "Although there is little reason to think that Augustine's position is authentically Christian... it nevertheless rapidly took hold and within an extremely short time had become universally accepted as fundamental Christian law" ([8], p. 71). "St. Augustine was the first to denounce suicide as a crime and thus shaped the later attitude of the Church regarding its sinfulness" ([16], p. 119).

The *third* reason for the comparative rarity of condemnations of suicide by the early church fathers is that suicide simply was not a problem for the early Christian community. In spite of the misconceived preconceptions and hence the erroneous, heuristic conclusions of some modern scholars who see the early Christian community as prone to suicide, there is absolutely no evidence in the corpus of Christianity for the first 250 years of the Christian era of any Christian under any circumstances commiting suicide for any reason. In the absence of even a shred of evidence of suicide by Christians occurring during this period, it is much more reasonable and responsible to assume that suicide did not present itself as a moral problem simply because it was so inherently contrary to Christian values and priorities as not to be considered a viable option for Christians than to conclude that, because suicide was not frequently condemned in the literature, it was not only regarded as appropriate but was even encouraged.

My purpose in the present essay is to argue the following: *First*, the New Testament, although it nowhere specifically condemns suicide, provides no implicit, much less explicit, encouragement of suicide. Rather it provides a structure of values and hopes antithetical, indeed inimical, to suicide. *Second*, the patristic attitude toward suicide is a quite consistent development of foundational issues of life and death, persecution and martyrdom, sanctification and suffering, and the sovereignty of God as expressed in the New Testament. Hence, *third*, the relevant material in patristic literature shows that it is misleading to suggest that Augustine formulated what then became the 'Christian position' on suicide. Rather, by removing certain ambiguities, he clarified and provided a theological cogent explanation of and justification for the position typically held by earlier and contemporary Christian sources. *Fourth*, Augustine's influence on later Christian thought and practice must be viewed circumspectly. It is ironic that the only category of suicide about which he disagreed with some of his contemporaries is that in which he continues to be at variance with some Catholic moral theologians. Furthermore, Augustine cannot reasonably be credited with any responsibility for those medieval (and much later) ecclesiastical and secular treatments of suicides that most of us today abhor.

The New Testament

The New Testament nowhere condemns suicide. Indeed, the act is never discussed. The only suicide recorded in the New Testament is Judas, and the event is simply reported without comment (Mt. 27:5; Acts 1:18), although it is reasonable to assume that he would hardly have been regarded as a model of Christian probity. Suicide comes up incidentally on a few other occasions. When Jesus said, "Where I am going, you cannot come", the Jews erroneously suggested that he might be contemplating suicide (Jn. 8:21–22). Neither condemnation nor approbation of the act may be read into this event. There is, however, one suicide prevented in the New Testament. When Paul and Silas were miraculously freed from prison in Philippi, the jailer was at the point of killing himself. As an alternative to suicide, Paul offered him salvation, which he joyfully accepted (Acts 16:25–34). Finally, self-destructive tendencies appear in some cases of demon possession, as in Mark's account of the Gerasene demoniac (Mk. 5:5) and of the mute boy (Mk. 9:14–29). In the latter instance, the boy's father told Jesus that from childhood the demon had "often cast him into the fire and into the water, to destroy him" (Mk. 9:21–22). This is the closest that the New Testament comes to the nebulous realm of suicidal psychoses. The New Testament contains no other references to suicide. It does, however, provide a cogent theological framework for the consideration of numerous moral issues not specifically addressed within its pages, including that of suicide.

There is considerable diversity in the New Testament. If one views that disparate collection of documents as a homogeneous whole without appreciating its varied strands of emphases, such as embodied in the synoptic gospels, the Johannine, Pauline, and Petrine literature, and such individual pieces as Acts, the Epistle to the Hebrews, and the letters of James and Jude, one fails to appreciate the richness of the unity of the New Testament that is highlighted by its diversity. The unity of precept in the various strands of New Testament thought is particularly striking when considering the issues surrounding suicide.

In the New Testament the words 'life' and 'death' are used both literally and metaphorically. When used literally, they refer to physical life and physical death. But when used metaphorically, and this is more frequently the case, they denote spiritual life and spiritual death.[4] Sometimes the words are used both literally and metaphorically in the same sentence, as when Jesus said, "I am the resurrection and the life; he who believes in me, though he die, yet shall he live, and whoever lives and believes in me shall never die" (Jn. 11:25–26).[5] Jesus was here assuring his followers that although they would experience physical death, they would never experience spiritual death.[6] The New Testament consistently juxtaposes the certainty of physical death and the certainty of eternal life. Jesus promised his followers, "when I go and prepare a place for you, I will come again and will take you to myself, that where I am you may be also" (Jn. 14:3). It is this assurance of the promise of resurrection, Christ's return, and impending heavenly glory,[7] that conditions the New Testament attitude toward physical life and physical death. Physical life, at its very best, is acknowledged as brief and uncertain,[8] and spiritual life is regarded as being of inestimably greater value. Jesus taught his disciples that "whoever would save his life will lose it; and whoever loses his life for my sake, he will save it. For what does it profit a man if he gains the whole world and loses or forfeits himself?" (Lk. 9:24–25 = Mt. 16:25–26).[9] In this sense, "he who loves his life loses it, and he who hates his life in this world will keep it for eternal life" (Jn. 12:25).

For Jesus' followers, this mortal life was simply to be a pilgrimage.[10] Hence it is not surprising that Jesus told his disciples not to "fear those who kill the body but cannot kill the soul; rather fear him who can destroy both soul and body in hell" (Mt. 10:28 = Lk. 12:4–5). He assured his disciples that no one could snatch them out of his hand (Jn. 10:28). Since Christ's followers belong to him,[11] they should not fear physical death. As Paul writes to the Romans, "If we live, we live to the Lord, and if we die, we die to the Lord; so then, whether we live or whether we die, we are the Lord's. For to this end Christ died and lived again, that he might be Lord both of the dead and of the living" (Rom. 14:8–9). Earlier in the same epistle he had asked, "Who shall separate us from the love of Christ?" (Rom. 8:35). His answer was that nothing can "separate us from the love of God in Christ Jesus our Lord" (Rom. 8:39), and the list of examples of realities that cannot separate the believer from Christ is headed by death (Rom. 8:38). Hence Paul, when writing to the Corinthians, can quote Isaiah 25:8, "Death is swallowed up in victory", and Hosea 13:14, "O death, where is thy victory: O death, where is thy sting?", and then go on to say, "The sting of death is sin, and the power of sin is the law. But thanks be to God, who gives us the victory through our Lord Jesus Christ" (1 Cor. 15:54–57).

The sting of death having been removed, the fear of death having been vitiated, Jesus' followers were to look to "an inheritance which is imperishable, undefiled, and unfading, kept in heaven" for them (1 Pet. 1:4), where their "names are written" (Lk. 10:20).[12] While some authors of the New Testament euphemistically refer to death as 'sleep',[13] for the Christian, death is also described as a state of being "at home with the Lord" (2 Cor. 5:8).[14] When Paul was in prison, not certain whether he would be freed or executed, he wrote to the church at Philippi, "For to me to live is Christ, and to die is gain" (Phil. 1:21). Life for Paul was to love and serve Christ. Hence death for him was gain: to have more of Christ. If the choice were Paul's, he would go to be with his Lord: "My desire is to depart and to be with Christ, for that is far better" (1:23). But his continued service he regarded as more necessary (1:24). Regardless, his supreme ambition was that "Christ will be honored in my body, whether by life or by death" (1:20).

Whether reading the Gospel narratives or the epistles, the pattern that one sees for the Christian's life is to do God's will. Jesus had told the crowds around him, "whoever does the will of my Father in heaven is my brother, and sister, and mother" (Mt. 12:50). This will be the mark of those admitted to God's kingdom: "Not every one who says to me, 'Lord, Lord', shall enter the kingdom of heaven, but he who does the will of my Father who is in heaven" (Mt. 7:21). God's will is expressed in Christ's commandments. Jesus said, "If you love me, you will keep my commandments" (Jn. 14:15).[15] His commandments are varied but are summed up in his exchange with a scribe who asked him, "Which is the great commandment in the Law?" Jesus replied: "You shall love the Lord your God with all your heart, and with all your soul, and with all your mind. This is the great and first commandment. And a second is like it, You shall love your neighbour as yourself. On these two commandments depend all the law and the prophets" (Mt. 22:34–40).[16]

Potential applications of these two commandments are enormously diverse, and specific injunctions contained in the New Testament are numerous; but all may be conveniently placed under two rubrics: service and holiness. Service to God and to man includes caring both for people's spiritual needs and for their material needs. Caring for their spiritual needs requires evangelism, which is bearing witness to the truth of the gospel. Holiness involves a sustained warfare against sin, that is, against one's own sins or propensity to sin. These categories provide the framework for a consideration of various types of, and motivations for, suicide. If any form of martyrdom can reasonably be regarded as suicide, then the New Testament teaching most specifically applicable to the question is on bearing witness for Christ. If instances of suicide from despair over the struggle within, from a desire to preserve holiness in the face of assaults against it, or if instances of death resulting from severe asceticism can be vindicated by reference to the New Testament, they would have to be considered within the context of its calls to holiness, as would suicide to escape the afflictions of life, when seen in the context of New Testament teaching about the edificatory and sanctifying role of suffering in the Christian's life.

The imperative to evangelize runs throughout the New Testament.[17] Evangelism, of course, is the attempt to proselytize by being a witness for Christ, that is, to spread the 'good news', the *euangelion*. 'Witness' in Greek is *martyr*, a word that came to take on a very important and highly specialized meaning within the first 150 years of the Christian era. Proclaiming the gospel often did provoke both resistance and persecution. But persecution was to be expected. Jesus had assured his disciples that if people persecuted him, they would also persecute them.[18] Indeed it was to be regarded as a special sign of blessedness when Christians were persecuted for righteousness' sake, for "theirs is the kingdom of heaven". Christians were to rejoice when reviled and persecuted for Christ's sake, "for your reward is great in heaven, for so men persecuted the prophets who were before you" (Mt. 5:10–12). Hence Peter and the apostles, after they had been whipped by order of the Sanhedrin and were released, went on their way "rejoicing that they were counted worthy to suffer dishonor for the name" (Acts 5:41). Paul writes to the Philippians that "it has been granted to you that for the sake of Christ you should not only believe in him but also suffer for his sake" (Phil. 1:29). The verb here translated "granted" is a form of *charidzomai*, a word that means 'to bestow graciously'. These sufferings, then, were not the exclusive privilege of a select few but of anyone who follows Christ in the manner Paul describes when he writes to Timothy, "Indeed all who desire to live a godly life in Christ Jesus will be persecuted" (2 Tim. 3:12).

Bearing witness to Christ was not optional. Jesus told his disciples, "every one who acknowledges me[19] before men, I also will acknowledge before my Father who is in heaven; but whoever denies me before men, I also will deny before my Father who is in heaven" (Mt. 10:32–33 = Lk. 12:8). Significantly, just a few sentences earlier, he had admonished them not to "fear those who kill the body but cannot kill the soul" (Mt. 10:28). The intervening instruction assures them of their value to God, who knows even the number of hairs on their heads. The Christian is obligated to confess or acknowledge Christ. Not to confess him is to deny him. Confessing Christ will provoke persecution. Persecution in its most violent or extreme forms may result in martyrdom.

Does the New Testament provide any encouragement for courting martyrdom? In a sense, yes. Jesus taught that one should rejoice when persecuted; the apostles felt that it was an honor when they were counted worthy of suffering for Christ's sake, and Paul saw persecution as a gift graciously given by God. Furthermore, in several passages the correspondence between the sufferings and the glory of Christ is stressed.[20] In suffering, the Christian was said to share or participate in Christ's suffering. Sometimes this was linked with the idea of sharing in Christ's glory. In his letter to the Romans Paul writes that Christians are heirs of God and co-heirs with Christ, "provided we suffer with him in order that we may also be glorified with him. I consider that the sufferings of this present time are not worth comparing with the glory that is to be revealed to us" (Rom. 8:16–18). In 1 Peter the same ideas are joined together: "But rejoice in so far as you share Christ's sufferings, that you may also rejoice and be glad when his glory is revealed" (1 Pet. 4:13).[21]

These ideas, however, cannot reasonably be taken as injunctions to seek martyrdom, when tempered by the precepts and examples that the New Testament provides for those who would bear witness for Christ. Paul urges his readers, "If possible, so far as it depends upon you, live peaceably with all" (Rom. 12:18). And James, having defined "the wisdom from above" as "pure, then peaceable, gentle, open to reason, full of mercy and good fruits, without uncertainty or insincerity", maintains that "the harvest of righteousness is sown in peace by those who make peace" (Ja. 3:17–18). Paul's "fruit of the Spirit" includes peace along with love, joy, patience, kindness, goodness, faithfulness, gentleness, and self-control (Gal. 5:22–23). Likewise, Peter admonishes his audience to have "a tender heart and a humble mind" (1 Pet. 3:8). Just a few sentences later he advises: "Always be prepared to make a defense to any one who calls you to account for the hope that is in you, yet do it with gentleness and reverence" (1 Pet. 3:15). Similarly, Paul says, "Let your speech always be gracious, seasoned with salt, so that you may know how you ought to answer every one" (Col. 4:6). These epistolary injunctions correspond very closely to the commands regarding conduct and the precepts of character that Jesus gave in his Sermon on the Mount (Mt. 5–7).

Furthermore, the Christian's actions while being persecuted for righteousness' sake were themselves to be a form of witness to non-Christians. Paul remarks that in his various sufferings he conducted himself in such a manner that his ministry would not be discredited (2 Cor. 6:3). Not only was the Christian to refrain from giving offense by his response to affliction, but he was to display a positive witness of love when persecuted. Christ says that although one commonly hears the adage, "You shall love your neighbor and hate your enemy", his teaching is, "Love your enemies and pray for those who persecute you" (Mt. 5:43–44). Paul, in the same spirit, writes to the Romans: "Bless those who persecute you; bless and do not curse them" (Rom. 12:14). Specifically, Christians were to follow Jesus as their example.[22] The Gospel narratives show that Jesus often withdrew from his enemies. Nor did he give himself up to the authorities in Jerusalem. Rather, he simply allowed himself to be arrested. Also he told his disciples that men would deliver them up (Mt. 10:17), not that they should deliver themselves up. Paul fled from Damascus to avoid those who wanted to kill him, and his flight was quite ignominious and personally humbling: he allowed himself to be lowered over the walls of the city in a basket (Acts 9:23–25).[23] And when he and Silas were persecuted in Thessalonica, their fellow Christians sent them away by night to a safer city (Acts 17:1–10).

Is avoiding or fleeing from persecution denying Christ? The New Testament answers this question with an unequivocal "no". Jesus thus instructed his disciples: "When they persecute you in one town, flee to the next" (Mt. 10:23). This warning is followed by the admonition not to fear those who can kill the body, and the obligation to confess Christ before men. Furthermore, Jesus taught his disciples to pray, "Lead us not into temptation" (Mt. 6:13 = Lk. 11:4). The word here translated 'temptation' (*peirasmos*) also is frequently rendered 'trial' or 'testing'. There is a tension in the New Testament between the repeatedly stressed reality of temptation, trial, testing in the Christian's life and the legitimized, indeed commanded, desire to avoid temptation, trial, testing – although the Christian is to rejoice when responding to various *peirasmoi* without sin. If there is no testing or trial of one's faith and no temptation to apostasy stronger than an imminent, painful death, the petition "Lead us not into temptation" should itself preclude a precipitous courting of martyrdom.[24] The New Testament clearly presents an obligation to die for Christ if the only alternative is to deny him. Significantly, the list of people to be given over to eternal punishment, that is, to the 'second death', is headed by the 'cowardly', followed by the 'faithless' (i.e., 'untrustworthy' or 'unbelieving', Rev. 21:8). On the other hand, there are, as we have seen, passages in the New Testament that not only discourage seeking martyrdom, but encourage, both by precept and by example, flight from persecution at least under some circumstances.

Let us now consider whether New Testament teaching on holiness encourages or permits 1) suicide owing to despair over the struggle with sin; 2) severe self-discipline that can lead to death; or 3) suicide to escape God's discipline.

Sin and holiness, diametrical opposites, are frequently contrasted in the New Testament. Christians are regularly reminded what their way of life was like when they were still "dead in their trespasses and sins". They are urged not to revert back to their old ways. Peter writes, "As obedient children, do not be conformed to the passions of your former ignorance, but as he who called you is holy, be holy yourselves in all your conduct; since it is written, "You shall be holy, for I am holy" (1 Pet. 1:14–16).[25] Spiritual life necessitates death to sin and to self (2 Cor. 4:11; 5:15), that is, conformity to the will of God which Paul describes as "your sanctification" (1 Thess. 4:3): "For God has not called us for uncleanness, but holiness. Therefore whoever disregards this, disregards not man but God, who gives his Holy Spirit to you" (1 Thess. 4:7–8). Hence, as the Epistle to the Hebrews exhorts, "Strive...for holiness without which no one will see the Lord" (Heb. 12:14). The authors of the New Testament never deny that such striving requires considerable effort. "For", as Paul writes, "the desires of the flesh are against the Spirit, and the desires of the Spirit are against the flesh; for those are opposed to each other, to prevent you from doing what you would [i.e., what you desire]" (Gal. 5:17). Paul's own frustration – frustration is too weak a word; anguish is better – is shown in his letter to the Romans: "I do not understand my own actions. For I do not do what I want, but I do the very thing I hate.... I know that nothing good dwells within me, that is, in my flesh. I can will what is right, but I cannot do it. For I do not do the good I want, but the evil I do not want is what I do.... [W]hen I want to do right, evil lies close at hand.... Wretched man that I am! Who will deliver me from this body of death?" (Rom. 7:15–24). Paul, of course, does not stop there, but goes on to rejoice that "there is now no condemnation for those who are in Christ Jesus" (Rom. 8:1). The New Testament authors do not deny either the potential for sin in the Christian's life or the reality of the struggle. But they also assure their readers of forgiveness and ultimate victory.[26] Hence despair over sin has no place in the New Testament that could justify, much less encourage, suicide. The New Testament consistently offers salvation to the non-Christian who despairs over sin; and it consistently promises assurance of restoration and progress in the spiritual life to the Christian who has fallen into sin[27] or is discouraged by the conflict that rages both within and without. To Christian and non-Christian alike, such a message is antithetical to encouraging suicide.

Overall, there is considerably more emphasis upon joy in believing and upon ultimate victory over sin in the New Testament than upon the necessity of the struggle against sin. This is not to say that the latter is not emphasized. It most certainly is, as even the few passages cited above should demonstrate. The struggle against sin is one aspect of the sanctifying process. It is the active part, that in which the Christian is expected to take the initiative. There is also a rather passive aspect, in which the Christian is expected merely to respond. The first of these is pertinent in a discussion of two of the types of, or motivations for, suicide that some modern authors attribute to early Christians. The second, as we shall see, is also relevant to a discussion of suicide. We can label the first of these self-discipline, the second God's discipline.

The self-discipline to which the Christian is called involves the striving mentioned above. Sometimes this is called 'mortification', that is, a 'putting to death', which is the literal meaning of the Greek word, *nekroō*, used in both Colossians 3:5 and Romans 8:13. The first of these reads, "Put to death therefore what is earthly in you: Immorality, impurity, passion, evil desire, and covetousness, which is idolatry". The second says, "if you live according to the flesh you will die, but if by the Spirit you put to death the deeds of the body you will live". Similarly Paul writes to the Galatians, "For he who sows to his own flesh will from the flesh reap corruption; but he who sows to the Spirit will from the Spirit reap eternal life" (Gal. 6:8). Earlier he had urged them, "[W]alk by the Spirit, and do not gratify the desires of the flesh" (Gal. 5:16). Paul says virtually the same thing to the church in Rome, when he tells them to "make no provision for the flesh, to gratify its desires". This is in conjunction with the command to "put on the Lord Jesus Christ" (Rom. 13:14). Peter likewise beseeches those to whom he was writing "as aliens and exiles to abstain from the passions of the flesh that wage war against your soul" (1 Pet. 2:11). These verses, along with Galatians 5:17 and Romans 7:15–24 quoted above, reveal a strong dichotomy between the 'flesh' and the 'spirit' or 'Spirit', and have, along with other statements made in the New Testament, sometimes provided the justification for an extreme denigration of the body, a severe asceticism that is based upon a radical dualism.

The Greek word *sarx*, usually translated 'flesh', has two distinct sets of meanings in the New Testament. The first is entirely neutral, having no moral thrust: the body itself, viewed as substance; the body with its physical limitations, the external, outward, physical side of life. The second ranges from mortal nature, that is, the state of mortality, which is essentially neutral in a moral sense, through human nature, that is, fallen human nature with its tendency earthward rather than Godward, ultimately to the conscious and willing instrument of sin. We have seen *sarx* used in this last, its most pejorative sense, several times in Paul's epistle to the Galatians. Yet, in that same letter, he writes, "I have been crucified with Christ; it is no longer I who live, but Christ who lives in me; and the life I now live in the flesh I live by faith in the Son of God, who loved me and gave himself for me" (Gal. 2:20). In his epistle to the Ephesians, after describing Christ's love and care for the church, which is called his body, Paul says, "Even so husbands should love their wives as their own bodies. He who loves his wife loves himself. For no man ever hates his own flesh, but nourishes and cherishes it, as Christ does the church, because we are members of his body" (Eph. 5:28–30). In such instances, *sarx* is used synonymously with *soma*, 'body', and just as neutrally. For instance, Paul's benediction to the Thessalonians includes his wish that their "spirit and soul and body be kept sound and blameless at the coming of our Lord Jesus Christ" (1 Thess. 5:23). Writing to the church in Corinth, in which some problems of sexual immorality had arisen, Paul asks, "Do you not know that your bodies are members of Christ?" (1 Cor. 6:15). After urging them to shun immorality, he asks, "Do you not know that your body is a temple of the Holy Spirit within you, which you have from God? You are not your own; you were bought with a price. So glorify God in your body" (1 Cor. 6:19–20).

"Glorify God in your body." This command, taken out of the broader moral context of New Testament teaching, can justify any use of the body that is construed by an individual as glorifying God. Paul, as we have seen, desired that "now as always Christ will be honored [literally 'glorified'] in my body, whether by life or by death" (Phil. 1:20). One who died for Christ certainly glorified God by his death. But glorifying God in one's body, other than by martyrdom, is an ambiguous concept. Obviously, one does not glorify God in one's body by breaking any of God's commandments, by sinning in any way; hence the command must be placed within the much broader framework of Christian ethics.

Let us now consider that form of self-discipline that sometimes leads to death, that is severe asceticism. Christian asceticism is motivated by a desire to subdue one's propensities to sin. Certain sects have now and then interpreted the doctrine of soteriology to require non-vicarious redemptive suffering, that is, earning favor with God through one's own propitiatory suffering. Such thought was sometimes bolstered by a perceived dualism between the spirit and the flesh, the latter being regarded as inherently evil. A basic principle of the New Testament, however, is that the Incarnation demonstrates that the material world was not rendered essentially evil by the Fall. The contrast between spirit and flesh presented in the New Testament is a vital dichotomy, not a dualism. The radical dualism that typified various groups in the early centuries of Christianity (e.g. Gnostics, Manicheans, Marcionites) that were declared heretical was founded upon a denigration of the entire material realm as essentially evil, leading to two opposite expressions: a sometimes lascivious antinomianism or a severe asceticism. But throughout the early centuries of Christianity there was a tension within orthodoxy between denying that the material realm was evil and desiring to subdue the flesh with its propensities to sin. Sometimes a functional, but not theologically radical, dualism strained the bounds of orthodoxy by exaggerating the *sarx-pneuma* dichotomy of the New Testament into a severe asceticism.

Does the New Testament support abuse of the physical body to any extent approaching a self-discipline that could cause death? We have already seen several examples of the numerous injunctions to spiritual self-discipline that appear within the New Testament. Furthermore, Jesus said to his disciples, "If your right eye causes you to sin, pluck it out and throw it away... And if your right hand causes you to sin, cut it off and throw it away". He concludes both statements thus: "... it is better that you lose one of your members than that your whole body be thrown into hell" (Mt. 5:29–30). And Paul frequently compares the Christian with the athlete who diligently seeks to win the prize through training, and says of himself, "I do not run aimlessly, I do not box as one beating the air; but I pommel my body and subdue it, lest after preaching to others I myself should be disqualified" (1 Cor. 9:26–27). Even during Paul's lifetime there were some who were advocating a radical (pre-Gnostic or early Gnostic) dualism. He wrote his epistle to the Colossians in part to counter such teaching: "These have indeed an appearance of wisdom in promoting rigor of devotion and self-abasement and severity to the body, but they are of no value in checking the indulgence of the flesh" (Col. 2:23).[28] The New Testament clearly asserts that it is the carnal mind – that is, "the mind that is set on the flesh" – which is "hostile to God". It "does not submit to God's law, indeed it cannot"; hence those who are in that sense "in the flesh ... cannot please God" (Rom. 8:7–8). Abusing one's body, which is described as the "temple of the Holy Spirit", profits nothing. It appears that the New Testament, on balance, does not encourage any self-discipline or asceticism that could be so severe as to cause one's death.

Let us turn from self-discipline to God's discipline, from self-imposed austerities to the sufferings and afflictions that are regarded as simply an inevitable part of the human condition. Life becomes exceedingly difficult sometimes owing to matters over which the individual has no ultimate control, matters that have traditionally been subsumed under the rubric of 'providence'. For the Christian, all circumstances that impinge upon and disturb his settled existence can be regarded (from a New Testament point of view) as discipline. The Epistle to the Hebrews teaches that Christ, although God's own Son, "learned obedience through what he suffered" (Heb. 5:8). Since Christ's divinity and absolute moral perfection are stressed in this epistle, we must assume that this learning of obedience through suffering is not intended as a demonstration of Christ's deficient obedience of which his sufferings were corrective. The passage seems to indicate that, in his incarnate state, he was in some way edified by his sufferings, for the next verse begins with the statement that by this he was made perfect, that is, complete. Earlier in the same epistle it is explicitly stated that "it was fitting that he [sc. God], for whom and by whom all things exist, in bringing many sons to glory, should make the pioneer of their salvation perfect through sufferings" (Heb. 2:10). Now if in New Testament thought, in some mysterious way, Christ himself, although never deficient in obedience, could be taught obedience through suffering; if he, although incarnate God, could be edified and perfected through suffering; so also then could suffering be a most beneficient and edificatory factor in the lives of his followers.

Just as being persecuted for righteousness is a consequence of living a godly life in Christ Jesus, so also is suffering in other forms part of God's edification of his children, according to the Epistle to the Hebrews. The passage in question includes a quotation from Proverbs: "My son, do not regard lightly the discipline of the Lord nor lose courage when you are punished ['reproved' is a better translation of *elencho* here] by him. For the Lord disciplines him whom he loves, and chastises every son whom he receives". The author of the Epistle to the Hebrews comments thus:

> It is for discipline that you have to endure. God is treating you as sons; for what son is there whom his father does not discipline? If you are left without discipline ... then you are illegitimate children and not sons. ...[God] disciplines us for our good, that we may share his holiness. For the moment all discipline seems painful rather than pleasant; later it yields the peaceful fruit of righteousness to those who have been trained by it (Heb. 12:5–11).

This passage tells us little about the nature of the discipline that God will visit upon his children. It is described as a reproving or a rebuking and as a whipping (the word translated "chastise" literally means 'scourge' or 'whip') and is certainly something that at the time is unpleasant, indeed painful. But discipline's effect on the Christian's character, if the discipline is properly received, is said to be salutary and more than sufficient compensation for the pain, since it bears a peaceful fruit of righteousness to those who have been trained by it.

Numerous passages can be adduced from the New Testament where the edificatory aspects of suffering are stressed. For example, in the epistle to the Romans, Paul writes:

> [W]e rejoice in our hope of sharing the glory of God. More than that we rejoice in our sufferings, knowing that suffering produces endurance, and endurance produces character and character produces hope, and hope does not disappoint us, because God's love has been poured into our hearts through the Holy Spirit which has been given to us (Rom. 5:2–5).

In 1 Peter we read, "do not be surprised at the fiery ordeal which comes upon you to prove you, as though something strange were happening to you" (1 Pet. 4:12). A complementary passage occurs in James: "Count it all joy when you meet various trials, for you know that the testing of your faith produces steadfastness. And let steadfastness have its full effect, that you may be perfect and complete, lacking in nothing" (Jas. 1:2–4).

It seems, then, that in New Testament terms, suffering for the Christian should be viewed as something in which he can rejoice, for two reasons: 1) The very fact that the Christian experiences, and responds as he does, to certain kinds of suffering, is proof of his sonship; and 2) God is using the suffering as a means of causing his child to grow spiritually, if the response to the affliction is right. Paul, writing to the Thessalonians, says,

> [W]e ourselves boast of you in the churches of God for your steadfastness and faith in all your persecutions and in the afflictions which you are enduring. This is evidence of the righteous judgment of God, that you may be worthy of the kingdom of God, for which you are suffering (2 Thess. 1:4–5).

Jesus, after telling the parable of the sower and the seeds, explained to his disciples the meanings of its various categories. Speaking of the seed sown on a rocky place as representing one who hears the word and immediately receives it with joy, he says that such a man "has no root in himself, but endures for a while and when tribulation or persecution arises on account of the word, immediately he falls away" (Mt. 13:20–21 = Mk. 4:16–17). Such a response to affliction and persecution stands in stark contrast to that of the Thessalonians and to the principle expressed in James: "Blessed is the man who endures trial, for when he has stood the test he will receive the crown of life which God has promised to those who loved him" (James 1:12). Perseverance, endurance, standing firm, overcoming – to such is the Christian exhorted. Christ had told his disciples that "he who endures to the end will be saved" (Mt. 10:22 = Mk. 13:13).[29] The same theme is emphasized in various epistles and especially in the book of Revelation.[30]

Does the New Testament permit or encourage despair in the face of affliction that would justify suicide? We already noted that the author of the Epistle to the Hebrews (12:5–6) urges his readers to avoid two extremes when under God's hand of discipline:[31] (1) making light of God's discipline; or (2) being crushed by, or losing heart or courage because of, his reproof. Believers are encouraged in their afflictions by such exhortations as the following which appears in Paul's second letter to the Corinthians, immediately after mention of the coming resurrection:

> So we do not lose heart. Though our outer nature is wasting away, our inner nature is being renewed every day. For this slight momentary affliction is preparing for us an eternal weight of glory beyond all comparison, because we look not to the things that are seen but to the things that are unseen; for the things that are seen are transcient, but the things that are unseen are eternal (2 Cor. 4:16–18).

This partakes of a theme dear to several New Testament authors: the troubles that must be faced here of small consequence in comparison with the joys of heaven.

Christians were to engage actively in a direct ministry to their fellow sufferers. For example, in the book of Acts, Paul and Barnabas are described as "strengthening the souls of the disciples, exhorting them to continue in the faith" in the face of afflictions (Acts 14:22). The word here translated 'exhort' also means 'encourage', 'comfort', or 'console', the same word that recurs in the following passage from 2 Corinthians:

> Blessed be the God and Father of our Lord Jesus Christ, the Father of mercies and God of all comfort, who comforts us in all our affliction, so that we may be able to comfort those who are in any affliction, with the comfort with which we ourselves are comforted by God. For as we share abundantly in Christ's sufferings, so through Christ we share abundantly in comfort too. If we are afflicted, it is for your comfort and salvation; if we are comforted, it is for your comfort, which you experience when you patiently endure the same sufferings that we suffer. Our hope for you is unshaken; for we know that as you share in our sufferings, you will also share in our comfort (2 Cor. 1:3–7).

There are several points that must be stressed here. In this passage the comfort is said to abound through Christ. He is the mediator through whom it comes. But it is to flow through the sufferer to other sufferers. Suffering thus is a training ground that enables Christians through participating in Christ's sufferings to participate in and minister to the sufferings of their fellow Christians. The relief that is provided is not the removal of the suffering but a consolation, a comfort in the suffering that transforms it into a positive force in the Christian's life when his response is positive.

A positive response to suffering is exemplified by Paul's attitude to what he calls his "thorn in (or for) the flesh":

> And to keep me from being too elated by the abundance of revelations, a thorn was given to me in the flesh, a messenger of Satan, to harass me. Three times I besought the Lord about this, that it should leave me; but he said to me, "My grace is sufficient for you, for my power is made perfect in weakness". I will all the more gladly boast of my weaknesses, that the power of Christ may rest upon me. For the sake of Christ, then, I am content with weaknesses, insults, hardships, persecutions, and calamities; for when I am weak, then I am strong (2 Cor. 12:7–10).

Here is an example of a central paradox in New Testament thought. Strength comes only through weakness. This strength is Christ's strength that comes only through dependence on him. And that is a child-like dependence. Christ had said, "[U]nless you turn and become like children, you will never enter the kingdom of heaven" (Mt. 18:3). He was not suggesting that his followers become as little children in selfishness or pride but in their dependence on God in all things.

In the Gospel of John Christ says: "I have said this to you, that in me you may have peace. In the world you have tribulation; but be of good cheer, I have overcome the world" (Jn. 16:33). "In the world you have tribulation." For Christians of the New Testament, it was simply to be expected and accepted. But no suffering was to be meaningless for them. The ultimate purpose and meaning behind the sufferings of Christians in the New Testament was their spiritual growth. And the ultimate goal in spiritual growth was a close dependence on God, a dependence that was based on a child-like trust. Numerous pasages can be adduced from the New Testament that encourage Christians to trust in their Heavenly Father. For example,"Have no anxiety about anything, but in everything by prayer and supplication with thanksgiving let your requests be made known to God. And the peace of God, which passes all understanding, will keep your hearts and your minds in Christ Jesus" (Phil. 4:6–7).[32]

The New Testament depicts Christians as children of their Heavenly Father who loves them and, in spite of any appearance to the contrary, is causing all things to work together for their good. Hence they should say with Paul,

> We know that in everything God works for good with those who love him, who are called according to his purpose... For I am sure that neither death, nor life, nor angels, nor principalities, nor things present, nor things to come, nor powers, nor height, nor depth, nor any other creatures will be able to separate us from the love of God which is in Christ Jesus our Lord (Rom. 8:28, 38–39).

To New Testament Christians, God was a *person* to whom they could relate in a spiritual depth commensurate with the degree of their commitment to the person of Christ. It was the intimacy of this relationship, a relationship based upon love and trust, that gave meaning to suffering for New Testament Christians and enabled Paul to say of himself and of those of kindred spirit, "we rejoice in our sufferings" (Rom. 5:3). Hence Peter, when writing to some Christians who were deeply afflicted, encouraged them to persevere. Instead of suggesting that they take their own lives, he says, "Let those who suffer according to God's will do right and entrust their souls to a faithful creator" (1 Pet. 4:19). Likewise, James writes, "As an example of suffering and patience, brethren, take the prophets who spoke in the name of the Lord" (Jas. 5:10). Early Christians would have been familiar with the sufferings of the prophets of the Old Testament. Moses in despair had prayed that God would slay him (Num. 11:15). So also had Elijah (1 Kg. 19:4) and Jonah (Jon. 4:3). Jeremiah cursed the day of his birth (Jer. 15:10, 20:14–18), as also did Job (Job 3:1). Job is the specific example of endurance that James gives: "Behold, we call those happy who were steadfast. You have heard of the steadfastness of Job, and you have seen the purpose of the Lord, how the Lord is compassionate and merciful" (Jas. 5:11). It is noteworthy that none of these Old Testament figures, however much they may have wished to depart from life, resorted to suicide. Nor is there any suggestion that they even considered it.[33] The early Christian would undoubtedly have recalled that Job's wife had urged him to curse God and die and that he rebuked her for such advice (Job 2:9–10). Furthermore, Job identified himself with those "who long for death, but it comes not" (Job 3:21) and says of himself, "I would choose strangling and death rather than my bones. I loathe my life" (Job 7:15–16). Job apparently did not regard suicide as an option even in the midst of his unbearable afflictions. Such, then, was the example of patience in suffering held up to early Christians.

In short, the New Testament encourages patience, stresses hope, commands perseverance, and so strongly emphasizes the sovereignty of God in the life of Christians and the trust that they should exercise, knowing that God will ultimately cause all things to work out for their good, that it is unlikely that suicide to escape from those very trials which God had chosen to inflict or permit could be regarded as anything less than a breach of trust.

The New Testament, however, does not deal with suicide by the mentally ill, for instance, by the severely depressed or melancholic. Several centuries would elapse before any extant Christian source would consider the theological or spiritual implications of such suicides.

THE PATRISTIC ERA

The Yale historian, George P. Fisher, in the second decade of this century, noted that for the Stoics, who justified suicide under many circumstances, "Life and Death are among the *adiaphora* – things indifferent, which may be chosen or rejected according to circumstances." He then remarked,

> How contrary is all this to the Christian feeling! The Christian believes in a Providence which makes all things work together for his good, and believes that there are no circumstances in which he is authorized to lay violent hands upon himself. There is no situation in which he cannot live with honor, and with advantage to himself as long as God chooses to continue him in being.

Fisher perspicaciously grasped the most essential values of early Christianity and as a consequence concluded, "Hence, in the Scriptures there is no express prohibition of suicide, and no need of one" ([20], pp. 174–175).

Fisher's assessment is also valid for the patristic ethos, with one modification: toward the end of the patristic era some sources did approve of one form of suicide, that is, by women to preserve their chastity. Were it not for the fact that patristic literature does include prohibitions of suicide, his conclusion would be equally accurate for post-New Testament Christianity, because the most basic Christian values expressed in the New Testament are the same values that undergird and are elaborated in patristic literature.

LIFE AND DEATH IN PATRISTIC THOUGHT

The church fathers expressed generally the same attitudes toward spiritual life and death, and physical life and death as the New Testament. Spiritual life was of infinitely greater value than physical life, and spiritual death was much more to be feared than physical death. Indeed, the Christian should not fear physical death at all, for it would simply be the means whereby he would be brought to those ineffable delights that heaven had in store for him. Numerous examples could be given from patristic sources, but typical is a treatise written by Ambrose (ca. 339–397) entitled *Death as a Good.* He begins by asserting that "Should death do injury to the soul, it can be considered an evil, but should it do the soul no harm, it cannot" (1.1 [19], vol. 65, p. 70). Only Christians have the correct perspective on life and death, and they have always "lamented the longevity of this pilgrimage, since they consider it more glorious 'to depart and to be with Christ'..." (*ibid.*, 2.3, p. 71). After an extensive discussion of a wide variety of related issues, he says, "To the just, death is a harbor of rest; to the guilty, it is reckoned a shipwreck" (*ibid.*, 8.31, p. 93). Then after quoting Col. 3:3–4, he begins his concluding paragraph with the exclamation, "let us therefore hasten to life" (*ibid.*, 12.57, p. 112).

Sentiments identical to those which Ambrose expressed can be found in virtually all the church fathers. Their attitude toward death is nicely described by Peter Brown when he observes that "the early church tended to leapfrog the grave. The long process of mourning and the slow adjustment to the great sadness of mortality tended to be repressed by a heady belief in the afterlife" ([10], pp. 69–70). Hence with great frequency we encounter statements such as the following by Chrysostom (349–407), "We should rejoice in the death of the righteous" (*Homily 8 on Phil.* [40], vol. 13, p. 233); and by Tertullian (ca. 160–ca. 220), "He who had gone ahead is not to be mourned, though certainly he will be missed". One's longing for the deceased was not to be a desire for the departed to be here but rather a desire to go and be with them. A few sentences later, Tertullian asserts that, "if we bear it with impatience and grief that others have attained their goal, we ourselves do not want to attain our goal" (*Patience* 9 [19] vol. 40, p. 209). This was frequently given as the mark of the truly committed and serious follower of Christ: a desire to die and be with Christ, demonstrated by a genuine envy for those who have gone 'home' already.

The same literature that consistently expressed a yearning for death also consistently expressed a respect for life. The *Shepherd of Hermas*, an anonymous work composed in stages between ca. 90 and ca. 150, asserts that one who is harassed by distress (*incommoda*) should be assisted, for "many bring death on themselves by reason of such calamities when they cannot bear them. Whoever therefore knows the distress of such a man, and does not rescue him, incurs great sin and becomes guilty of his blood" ("Similitude"

10.4.3.[4], vol. 2, p. 305). This passage suggests that the author regarded the suicide of one who resorted to it owing to distress as so serious that anyone who could have helped him but failed to do so not only had committed a serious sin but was also guilty of his blood. Early Christians regarded physical life as a gift of God that was so precious that they viewed the care of the sick as a categorical imperative. The gospel as proclaimed in the early centuries of Christianity did not limit itself to the salvation of souls for eternity, but was also directed to salvation within the world. The care of the destitute generally, and particularly of the sick, became a duty incumbent on all believers. Adolf Harnack writes regarding the obligation to visit and care for the sick that "to quote passages would be superfluous, for the duty is repeatedly inculcated" ([24], pp. 69–70). Early Christian literature is indeed rife with such admonitions.[34]

In spite of the sometimes extreme ascetic tendencies of the early church, a central tenet of Christian orthodoxy consistently confirms the inherent worth of life and the moral neutrality of the body. Clement of Alexandria (ca. 155 – ca. 220), for example, maintains that those who "vilify the body are wrong.... The soul of man is confessedly the better part of man, and the body the inferior. But neither is the soul good by nature, nor, on the other hand, is the body bad by nature" (*Stromateis* 4.26 [3], vol. 2, p. 439). He regarded health as a gift and insisted that the body, as the temple of the Holy Spirit, deserved reasonable care. Clement approvingly quotes Plato's injunction that care must be taken of the body (*ibid.*, 4.4, p. 412). Tertullian, who can hardly be regarded as effete, says "...I do bathe at the hour I should, one which is conducive to health and which protects both my temperature and my life's blood" (*Apology* 42.4 [19], vol. 10, p. 107). The church fathers saw health as a potential good or evil depending on the Christian's use of it. They also saw sickness as a potential good.

Although they thought it proper to desire and to seek the restoration of health when ill, the church fathers regarded excessive concern for the body and a desperate clinging to life as a sad contradiction of Christian values. Cyprian (ca 200–210 – 258), writing in Carthage to his fellow Christians while the city was being beseiged by plague, was disturbed both by their fear of death and by their efforts to preserve their lives:

What madness it is to love the afflictions, and punishments, and tears of the world and not rather to hurry to the joy which can never be taken from us.... How absurd it is and how perverse that, while we ask that the will of God be done, when God calls us and summons us from this world, we do not at once obey the command of His will. We struggle in opposition and resist, and in the manner of obstinate slaves we are brought with sadness and grief to the sight of God, departing from there under the bond of necessity, not in obedience to our will.... Why, then, do we pray and entreat that the kingdom of heaven may come, if earthly captivity delights us? ...let us be ready for every manifestation of God's will; freed from the terror of death, let us think of the immortality which follows (*Mortality* 5, 18, and 24 [19], vol. 36, pp. 202–203, 214, and 219).

Basil of Caesarea (ca. 329–379), instructing his monks about the proper use of the medical art, was governed by principles similar to those that motivated Cyprian. Basil writes, "Whatever requires an undue amount of thought or trouble or involves a large expenditure of effort and causes our whole life to revolve, as it were, around solicitude for the flesh must be avoided by Christians" (*The Long Rule* 55 [19], vol. 9, p. 331).

In a sermon, John Chrysostom describes a woman who was urged by her Christian friends to employ supposedly efficacious but magical means for the cure of her critically ill child. Chrysostom praises her refusal to resort to such illicit means, even though she thought they would restore her child to health. He then laments to his audience the low level of spiritual life and the skewed priorities of so many professing Christians who are little concerned with heaven, although they are willing to undergo anything for the sake of this life. He urges his audience to be ready for death and asks them why they cling to the present life (*Homily 8 on Col.*). Similarly, Augustine (354–430) preached that just as the martyrs, even though they loved life, did not cling to it but willingly gave it up when God chose to remove them (*Sermon*, 345.2), so also should those afflicted with seemingly hopeless illness. He points to the irony that so many, when faced with troubles, cry out, "'O God, send me death; hasten my days'. And when sickness comes they hasten to the physician, promising him money and rewards" (*Sermon* 84.1 [23], p. 24–25). Augustine was grieved at

what things men do that they may live a few days.... If, on account of bodily disease, they should come into the hands of the physician and their health should be despaired of by all who examine them; if some physician capable of curing them should free them from this desperate state, how much do they promise? How much is given for an altogether uncertain result? To live a little while now, they will give up the sustenance of life (*Sermon* 344.5, *ibid.* p. 25).

Physical life was worth little to many early Christians. But it was also of inestimable value. The Christian was frequently urged not only to give his life willingly as a martyr if the only alternative was denying Christ, but when sick, although he should seek healing, whether miraculous or medical or both, he should not cling to life but to regard his sickness as potentially the

God–given vehicle for his 'homegoing'. And, under all circumstances, the care of the soul was to take precedence over the care of the body.

PERSECUTION AND MARTYRDOM

The subject of martyrdom in the early church is complex.[35] I cannot deal with the question of why Christians were persecuted but must simply consider reactions to persecution by those who were persecuted or in danger of death. Reactions to persecution took one of four responses: 1) denying Christ (apostatizing); 2) fleeing possible martyrdom; 3) accepting martyrdom when the only escape was denying Christ (apostatizing); and 4) seeking, provoking, or volunteering for martyrdom.

The third of these was always approved and the first was always condemned by those sources that represent orthodoxy during the patristic era. The major problem posed by apostasy in the face of persecution was whether the Christian community should receive apostates back into fellowship. That controversy, however, lies outside the purview of the present study. When I say that apostasy was consistently condemned by the orthodox community, I am excluding certain heretical groups that were ostensibly Christian. A good example are the Gnostics.

The Gnostic teacher Basilides of Alexandria (early second century) maintained that apostasy – even a light-hearted denial of Christ – was permissible if it would save one's life. A later Gnostic, Heracleon (late second century), taught that what one confessed with one's tongue before men was irrelevant to the condition of one's heart, which only God knows. The reaction of the Christian community to this position was strongly negative. Clement of Alexandria, for example, was totally unambiguous in his denunciation of the Gnostic position:

> Now some of the heretics who have misunderstood the Lord, have at once an impious and cowardly love of life; saying that the true martyrdom is the knowledge of the only true God (which we also admit), and that the man is a self-murderer and a suicide who makes confession by death; and adding other similar sophisms of cowardice (*Stromateis* 4.4 [3], vol. 2, p. 412).

Marshalling other examples would be superfluous: the orthodox community's condemnation of apostasy to escape persecution was unequivocal. Much more troublesome to the early Christian community were the second and fourth responses: physical flight from persecution to avoid death; and seeking, provoking, or volunteering for martyrdom. Both, of course, are closely related, since it can be and has been argued that refusing the former is tantamount to doing the latter.

Immediately after his condemnation of the Gnostics' glib attitude toward apostasy, Clement writes:

Now we, too, say that those who have rushed on death (for there are some, not belonging to us, but sharing the name merely, who are in haste to give themselves up, the poor wretches dying through hatred to the Creator) – these, we say, banish themselves without being martyrs, even though they are punished publicly. For they do not preserve the characteristic mark of believing martyrdom, inasmuch as they have not known the only true God but give themselves up to a vain death, as the Gymnosophists of the Indians to useless fire (*ibid.*).

A few chapters later, he picks up this subject again:

When, again, He [*sc.* Christ] says, "When they persecute you in this city, flee ye to the other", He does not advise flight, as if persecution were an evil thing; nor does He enjoin them by flight to avoid death, as if in dread of it, but wishes us neither to be the authors nor abettors of any evil to any one, either to ourselves or the persecutor and murderer. For He, in a way, bids us take care of ourselves. But he who disobeys is rash and foolhardy. If he who kills a man of God sins against God, he also who presents himself before the judgment-seat becomes guilty of his death. And such is also the case with him who does not avoid persecution, but out of daring presents himself for capture. Such a one, as far as in him lies, becomes an accomplice in the crime of the persecutor. And if he also uses provocation, he is wholly guilty, challenging the wild beast. And similarly, if he afford any cause for conflict or punishment, or retribution or enmity, he gives occasion for persecution (*ibid.*, 4.10, p. 423).

We must not suppose that Clement did not hold martyrdom in high esteem and regard it as an obligation if the alternative was apostasy. He writes that the true Christian "when called, obeys easily, and gives up his body to him who asks.... [I]n love to the Lord he will most gladly depart from this life" (*ibid.*, 4.4, p. 411). But we do see him adamantly opposed to seeking martyrdom. He demonstrated this attitude in practice by fleeing from Alexandria when persecution struck there.

But not all Christians shared his views. Indeed Clement's disciple and successor, Origen (ca. 185 – ca. 254), as a youth whose father was about to be martyred, wished to present himself to the magistrates for martyrdom. His plans were thwarted by his Christian mother, who hid his clothes to keep him home until the crisis passed. Clement's contemporary, Tertullian, in his *Scorpiace*, written while he was still a Catholic, speaks of the faithful being hunted down like rabbits (1.11). The imagery implies that he regarded flight as legitimate. In his treatise, *Patience*, written about the same time, he specifies that patience is tested by torture, martyrdom, and by the inconveniences of flight (13.6). But in his *Ad Uxorem*, also written during the same period, he asserts that although flight is permitted as preferable to apostasy, it is not good (1.3). This is the first hint of a peculiarity in Tertullian's thought at a time when he was drawing nearer to leaving the Catholic fold for Montanism. After his change of allegiance to this rigorist sect, he wrote his *Flight in Time of Persecution*, in which he denounced, in no uncertain terms, flight from persecution as a denial of Christ and a sign of cowardice. He explains Jesus' command to his disciples to flee when persecuted as applying only to the first generation of Christians, "For, if they had been killed right at the beginning, the diffusion of the Gospel, too, would have been prevented" (6.3 [19], vol. 40, p. 287).

We must not conclude from this, however, that Tertullian's final, negative position on the question of fleeing from persecution represents simply an extreme position peculiar to Montanism. Others within Catholicism regarded flight from persecution as tantamount to apostasy, or to bribery to which a fair number had resource, hence avoiding martyrdom without having to deny Christ by word or by performing pagan rituals. Tertullian, especially as a Montanist, represents a rigorist, Clement a moderate, position on the question. The debate continued well beyond their time as is well illustrated by the case of Peter, Bishop of Alexandria, in the early fourth century. During the Great Persecution (303–312), Peter fled from Alexandria to avoid martyrdom. After the persecution subsided, he returned and composed what may originally have been merely a paschal letter, but which was later incorporated into the canon law of the Greek Orthodox Church. In this letter he disparaged those who had volunteered for martyrdom and approved those who had saved their lives by bribery or by flight.[36] Peter's position certainly is not rigorist and is somewhat more liberal than even Clement's with whom he shares the opinion that the Christian who does not attempt to escape martyrdom shares moral culpability with the persecutors.

Numerous examples of the contrast between the rigorist and moderate positions could be introduced here. But let it be sufficient to direct the reader to the accounts of the martyrdom of Ignatius (martyred either in 98 or 117), on the one hand, and Polycarp (martyred between 155 and 160), on the other. In the letters that he wrote while awaiting execution, Ignatius displayed, as W.H.C. Frend says, "a state of exaltation bordering on mania" ([22], p. 197). In his letter *To the Romans*, Ignatius wrote:

I am writing to all the Churches, and I give injunctions to all men, that I am dying willingly for God's sake, if you do not hinder it. I beseech you, be not "an unseasonable kindness" to me. Suffer me to be eaten by the beasts, through whom I can attain to God. I am God's wheat, and I am ground by the teeth of wild beasts that I may be found pure bread of Christ. Rather entice the wild beasts that they may become my tomb, and leave no trace of my body, that when I fall asleep I be not burdensome to any. Then shall I be truly a disciple of Jesus Christ, when the world shall not even see my body. Beseech Christ on my behalf, that I may be found a sacrifice through these instruments. I do not order you as did Peter and Paul; they were Apostles, I am a convict; they were free, I am even until now a slave. But if I suffer I shall be Jesus Christ's freedman, and in him I shall rise free. Now I am learning in my bonds to give up all desires (4[4], vol. 1, p. 231).

His was a case of voluntary martyrdom. Polycarp presents quite a different picture. Yielding to the entreaties of his friends, he withdrew from his city to avoid arrest. Ultimately he was apprehended and, when suffering martyrdom, evinced a moving serenity in the face of death which is quite in contrast with Ignatius' passionate desire for martyrdom (Eusebius, *Ecclesiastical History* 4.15).

Both Eusebius (*ibid.*) and the anonymous, but contemporary, author of the *Martyrdom of Polycarp* record the example of a certain Phrygian named Quintus

who had forced himself and some others to come forward of their own accord. Him the Pro–Consul persuaded with many entreaties to take the oath and offer sacrifice. For this reason, therefore, brethren, we do not commend those who give themselves up, since the Gospel does not give this teaching (*Martyrdom of Polycarp* 4 [4], vol. 2, p. 317).

Clement and Peter of Alexandria would agree with this last sentence. Tertullian would not. In his last extant work, addressed to a Roman official named Scapula, Tertullian writes concerning persecution and martyrdom, that Christians, in this case, Montanists,

do not fear these things, but willingly call them down upon ourselves. When Arrius Antoninus [governor of the province of Asia ca. 184–185] was carrying out a vehement persecution in Asia, all the Christians of the city appeared in a body before his tribunal. After ordering a few to be led away to execution, he said to the rest, "Wretched men, if you wish to die, you have precipices and ropes to hang yourselves". If it should come into our mind to do the same thing here, also, what will you do with so many thousands of human beings ... giving themselves up to you? (*To Scapula* 5 [19], vol. 10, p. 160).

It may very well be that Arrius Antoninus missed the point that this group of Christians was trying to make. Christians were frequently frustrated with the government's inconsistencies in its treatment of Christians. Only spasmodically were Christians persecuted at the initiative of the government (as distinct from mob action and private accusations). Hence their frustration: "Either clarify your policy and apply it consistently or allow us to live in peace" (see, e.g., Tertullian, *Apology* 5). Aside from the incident described by Tertullian in his letter to Scapula, there is little evidence of groups of Christians presenting themselves en masse before officials.[37] But when it did happen (which was apparently extremely rare) or when individuals or groups volunteered for martyrdom, it undoubtedly smacked of theatrics.

In his invective sketch of Peregrinus, the profligate-turned-Christian-turned-Cynic, Lucian, a second-century Greek satirist, says about Christians that "the poor devils have convinced themselves they're all going to be immortal and live forever, which makes most of them take death lightly and voluntarily give themselves up to it" (*The Death of Peregrinus* 13 [27], p. 369). Although one need take satirists with a grain of salt, Lucian's assessment is probably not a significant exaggeration of the sentiments of many pagans who may have regarded Christians as suicidal for their willingness to be martyred. But such pagans, including Lucian and the emperor Marcus Aurelius, who regarded Christians as morbid exhibitionists (*Meditations* 11.3), would probably have been unaware of how many Christians did in fact unobtrusively flee from persecution to avoid martyrdom. Very likely the majority of Christians who were martyred accepted death "voluntarily" when the only alternative was apostasy. Dying "voluntarily" does not necessarily mean seeking martyrdom. We have already seen that Clement condemns unequivocally those who voluntarily give themselves up to the officials. But he is perfectly consistent when he says that the true Christian "will not forsake his creed through fear of death ... in love to the Lord he will most gladly depart from his life .... With good courage, then he goes to the Lord, his friend, for whom he voluntarily gives his body..." (*Stromateis* 4.4 [3], vol. 2, p. 411).

Boniface Ramsey asserts that, in patristic thought, "Since martyrdom was a charism, a grace, it could not be demanded as a right; it was a free gift of God" ([38], p. 126). Clement writes that true Christians are "distinguished from others that are called martyrs, inasmuch as some furnish occasions for themselves, and rush into the heart of dangers...." By contrast, true Christians, "in accordance with right reason, protect themselves", but, with "God really calling them, promptly surrender themselves, and confirm the call, from being conscious of no precipitancy" (*Stromateis* 7.11 [3], vol. 2, pp. 541–542). Hence, one can be certain that God is calling one to martyrdom only if one has done nothing to precipitate it. During an outbreak of plague in

Carthage many Christians were distressed because, if they died of the pestilence, they would be deprived of the possibility of martyrdom. Cyprian, in addressing their concern, maintained that "martyrdom is not in your power but in the giving of God, and you cannot say that you have lost what you do not know whether you deserved to receive" (*Mortality* 17 [19], vol. 36, p. 213).

The significance of martyrdom varied according to one's soteriology. To some, martyrdom was the only sure means of salvation. Tertullian, in his later years, is representative of his position. To others, martyrdom was one of several means of sanctification leading to salvation. We see this idea heartily emphasized by Clement. Those who held the former position would probably crave martyrdom more than the latter, even though their desire for martyrdom would not necessarily cause them actively to seek it. But even for Clement, martyrdom was the most perfect display of love and was to be desired above any other form of death. Other forms of death could never offer the spiritual glory that martyrdom provided. Hence for those who ardently wished to depart from this life, any form of death, including suicide, would be an obstacle to that one cherished form of death, martyrdom. Accordingly, those who most wished to die would seek martyrdom if their theology permitted it. Likely, the majority of those who wished to quit life held the position most commonly encountered in the literature, namely, that seeking martyrdom was wrong. The very basis for a condemnation of actively seeking that one laudable form of death would *eo ipso* preclude intentionally ending one's own life through some lesser means. It is likely that the vast majority of Christians before the legalization of Christianity in 313 not only did not seek martyrdom but held that Christ's admonition to flee when persecuted had an abiding validity. Those who believed that one should seek by flight to avoid the most glorious and spiritually fulfilling form of death would be very hard pressed to formulate a theological justification for actively seeking to end their lives by their own hand. It is not surprising that in the literature extant from the period before the legalization of Christianity, there is absolutely no mention of Christians, who could not succeed in provoking pagans to put them to death, committing suicide.

SUFFERING, SANCTIFICATION AND THE SOVEREIGNTY OF GOD

How did the church fathers feel that Christians should regard forms of suffering other than persecution? Were they to inflict suffering upon themselves in pursuit of sanctification? And what of those afflictions that beset humanity generally? Is God sovereign, and if so, how should Christians' appreciation of God's sovereignty affect their understanding of sanctification and, consequently, their response to tribulation? We shall begin by considering self-inflicted suffering.

Asceticism, which is the practice of strict self–denial as a spiritual discipline, was a marked feature of early Christianity beginning in the late second or early third century. Clement of Alexandria was the first Christian author to emphasize asceticism as an ideal on the same level as martyrdom. Later, when Christianity became a licit religion and Christians ceased to be martyred for their faith, the ascetic replaced the martyr in the minds of many Christians as the new spiritual hero. The ascetic way of life in its more extreme forms involved a vilification and abuse of the body by those who had withdrawn from society to engage in a determined effort to subdue indwelling sin. Their self–discipline was regarded as a 'daily martyrdom'.

As we have already seen, mortification of the flesh (i.e., the carnal mind) and denial of self are stressed in the New Testament. But the asceticism that developed in the late second and early third centuries and gained considerable momentum in the fourth went beyond a simple application of New Testament principles. The climate in which Christian asceticism arose was one in which various classical schools of philosophy extolled simplicity and frugality and in which some pagans yielded to the impulse to experience the "flight of the alone to the alone" and withdrew from society. A few of these became severe ascetics who sought suffering for expiatory, propitiatory, or purificatory ends by abusing their bodies as the prisons of their souls.[38] They were similar in some of their excesses to members of certain heretical groups, such as the Gnostics, Manicheans, and Marcionites (dualists who conceived of matter, including the body, as inherently evil). While the New Testament, particularly Paul, does speak of a dichotomy between flesh and spirit, some Christian ascetics strained the bounds of orthodoxy when they exaggerated this dichotomy by abusing the body for the good of the soul. During these centuries there was only a very fine line between a still orthodox but extreme mortification of the flesh and a dualistic, heretical denunciation of the flesh as inherently evil. Although many church fathers did vilify the body, for the most part their appraisal of the body's worth was tempered by their conviction that it was morally neutral, potentially either a temple or a tomb, and must be subservient to the soul in its latter's campaign against evil.[39] Augustine, who is typically regarded as exemplifying the orthodox spirit in this as in most matters, held that the body is but the slave of the soul. Hence abusing the body accomplishes nothing:

Now it may be asserted that the flesh is the cause of every kind of moral failing, on the ground that the bad behavior of the soul is due to the influence of the flesh. [But] those who imagine that all the ills of the soul derive from the body are mistaken. [For it is] not by the possession of flesh, which the Devil does not possess, that man has become like the Devil: it is by living by the rule of self, that is by the rule of man (*City of God* 14.3 [5], pp. 550–552).[40]

Much that the church fathers say about the body strikes our analgesic ethos as severe. Much of it also appears to contradict their stress on the value of life and on health as a good. For the modern audience it is one of the most enigmatic features of Christianity during the third, fourth, and fifth centuries. All the church fathers whom we have considered thus far regarded most things and most conditions as potentially good or as potentially evil, depending on the Christian's use of, or response to, them. Health could be a good thing, or it could be a bad thing. Likewise with sickness or with any other form or source of suffering. Jerome (ca. 345 – ca. 419), for example, writes to a remarkably healthy centenarian that the health of the righteous is God's gift in which they should rejoice, but the health of the unrighteous is Satan's gift to lead them to sin (*Letter* 10). Hence Jerome tells another correspondent to rejoice not only in health but also in sickness, saying, "Am I good in health? I thank my Creator. Am I sick? In this case also I praise God's will. For 'When I am weak, then am I strong', and the strength of the spirit is made perfect in the weakness of the flesh" (*Letter* 39.2 [41], vol. 6, p. 50). This is a subject to which Jerome's heart warms: sickness can cause people to adjust their priorities. He describes a young lady who was taught by a burning fever, with which she had suffered for nearly thirty days, to direct her attention to more serious pursuits than the pampering of her person, to which, apparently, she had been giving more time than Jerome thought appropriate (*Letter* 38.2). Ambrose likewise suggests to a correspondent who had been sick that God had sent the sickness to him for the sake of his spiritual health (*Letter* 79). And in his treatise *Concerning Repentance* he writes that while sickness restrains one from sin, luxury is a catalyst to sins of the flesh (1.13.63).

How was the Christian to respond when ill? We have already seen that he was encouraged to seek healing but admonished not to cling desperately to life. But was he to aggravate the condition, thereby increasing his suffering and thus supposedly derive some spiritual benefit? All the church fathers insisted on a practical self-denial and a subduing of the flesh. The extremes of asceticism of which they approved varied. Clement was very mild in that regard compared to Jerome who, with a hearty fortitude that well matched his often caustic personality, bordered on the severe in the self-denial that he both practiced and preached. In one letter Jerome writes that when one's limbs are weak from fasting, Satan may oppress him with illness. What then should he do? Why, respond to the devil just as Paul did, saying "When I am weak, then am I strong" and "Power is made perfect in weakness" (*Letter* 3.5).

On the other hand, in a letter to the lady Paula whose twenty-year-old daughter, Blaesilla, had recently died, Jerome chides her for fasting. He imagines Christ saying to her that such fasting, which simply gratifies her grief, is displeasing to him. "Such fasts are my enemies. I receive no soul which forsakes the body against my will." It would be suicide for her to die in this way. Interestingly, Blaesilla's death had probably been hastened by her severe self-abasement, which Jerome had then most heartily encouraged; he still approved of it when he wrote to Paula. Toward the end of the letter, he rebukes her for her public display of grief during the funeral procession. Not only is it a bad witness since Christians are supposed to rejoice in the 'homegoing' of loved ones, but also he knows what the Roman crowd viewing the procession were probably saying: The girl was "killed with fasting". "How long must we refrain from driving these detestable monks out of Rome?" "They have misled this unhappy lady… " (*Letter* 39 [41], vol. 6, p. 53). Apparently the fasting in which the already sick Blaesilla had engaged, which very likely contributed to her death, was, in Jerome's opinion, pleasing to Christ. But Paula's fasting was displeasing to Christ because her motivations were wrong. And if she had died because of it, it would have been suicide and consequently Christ would not have received her soul since it would have forsaken her body against his will.

There are some very fine lines to draw here. And we can rest assured that Jerome drew them very close to those extremes of asceticism that earned for the rigorist Messalians, as an example, the status of heretics by official proclamation of the orthodox community. Most of the church fathers would not have gone as far as Jerome did in exhorting others to persevere in subduing and denying the flesh. But all shared the view that the soul was of infinitely greater value than the body. They did, however, as we have seen, espouse an obligation to care for the body. Hence the tension between these two obligations as perceived by the Christian community at that time and for a long time after as well.

Leaving persecution, martyrdom, and asceticism, we turn to those forms of suffering that are the common lot of mankind. Even a cursory and random reading of the church fathers reveals that they regarded suffering as an essential aspect of God's sanctifying of his people. This belief, combined with a firm assurance that God is sovereign, and an equally firm trust that he does all thing for their ultimate good, engendered in them an imperative to preach and practice endurance in the face of all afflictions. Cyprian and Tertullian, who wrote in Latin, and Clement of Alexandria, who wrote in Greek, lived when Christians were subject to persecution and possible martyrdom. They differ from each other significantly in their backgrounds, personalities, and emphases and hence they reflect the diversity of theology that then prevailed within the realm of orthodoxy. Yet, in the most foundational and essential areas of Christian values, they display a profound unity: Christians are subject to the whole range of afflictions that beset fallen humanity in a fallen world. Indeed, Christians must face even greater sufferings than pagans, because both God and Satan will buffet them. Satan does so in order to discourage them and hence to tempt them to sin; God does so as paternal chastening (including both training and discipline) that leads to sanctification. Accordingly the Christian must practice patient endurance in defiance of Satan and in resignation to the salutary providence of God. It should be noted that both the Greek and Latin words typically translated as 'patient' have the underlying meaning of 'patient endurance'.

Cyprian was the most pastoral of these three church fathers. He also most clearly represents the mainstream of orthodoxy. Two of his most pastoral writings are *Mortality*, written during a time of plague, and *The Good of Patience*. In the latter he asserts that "a crown for sorrow and suffering cannot be obtained unless patience in sorrow and suffering precede" (*The Good of Patience*, 10 [19], vol. 36, p. 273). For with patience

> we may endure all afflictions.... It is a salutary precept of our Lord and Master: ' He who has endured even to the end will be saved' .... We must endure and persevere ... so that, having been admitted to the hope of truth and liberty, we can finally attain that same truth and liberty, because the very fact that we are Christians is a source of faith and hope. However, in order that hope and faith may reach their fruition, there is need of patience.

After quoting Romans 8:24–25, he says, "Patient waiting is necessary that we may fulfill what we have begun to be, and through God's help, that we may obtain what we hope for and believe." He quotes Galatians 6:9–10 and comments that Paul here "warns lest anyone, through lack of patience grow tired in his good work; lest anyone either diverted or overcome by temptation should stop in the middle of his course of praise and glory and his past works be lost ...." After quoting Ezekiel 33:12 and Revelation 3:11, he says that "these words urge patient and resolute perseverance, so that he who strives for a crown, now with praise already near, may be crowned because his patience endures". He immediately asserts that patience "not only preserves what is good, but also repels what is evil", for "it struggles ... against the acts of the flesh and the body whereby the soul is stormed and captured". He then enumerates various sins against which patience is the only efficacious defense and states that, if patience is strong, "the hand that has held the Eucharist will not be sullied by the blood-stained sword" (*ibid.*, 12–14, pp. 275–277). Later he maintains that patience is

> necessary in respect to various hardships of the flesh and frequent and cruel torments of the body by which the human race is daily wearied and oppressed. ... it is necessary to keep struggling and contending in this state of bodily weakness and infirmity; and this struggle and strife can not be endured without the strength of patience. But different kinds of sufferings are imposed on us to test and prove us, and many forms of temptations are inflicted upon us by loss of wealth, burning fevers, torments of wounds, by the death of dear ones. ... the just man is proved by patience, as it is written 'In thy sorrow endure and in thy humiliation keep patience, for gold and silver are tried in the fire' [Ecclesiasticus 2:4–5].

He then gives the example of Job: "Thus Job was examined and proved and raised to the pinnacle of praise because of the virtue of patience". He describes Job's various afflictions and says that,

> lest anything at all might remain which Job had not experienced in his trials, the devil even armed his wife against him.... Nevertheless, Job was not broken by these heavy and continuous assaults, and in spite of these trials and afflictions he extolled the praise of God by his victorious patience (*ibid.*, 17–18, pp. 279–281).

A little later he exclaims:

> [L]et us ... maintain the patience through which we abide in Christ and with Christ are able to come to God.... It is patience that both commends us to God and saves us for God.... It vanquishes temptations, sustains persecutions, endures sufferings and martyrdoms to the end. It is this patience which strongly fortifies the foundations of our faith. It is this patience which sublimely promotes the growth of hope (*ibid.* 20, pp. 282–283).

As the climax of his argument, he exhorts, "Let us ... persevere and let us labor ... watchful with all our heart and steadfast even to total resignation ..." (*ibid.* 24, p. 287).

In his treatise entitled *Mortality* Cyprian comments on the phenomenon that some Christians were troubled because this

> disease carries off our people equally with the pagans, as if a Christian believes to this end, that, free from contact with evils, he may happily enjoy the world and this life, and, without having endured all adversities here, may be preserved for future happiness.... But what in this world do we not have in common with others as long as

this flesh ... still remains common to us?

He gives as examples famine, the ravages of war, drought, shipwreck; "and eye trouble and attacks of fever and every ailment of the members we have in common with others as long as this common flesh is borne in the world" *(Mortality* 8, [19], vol. 36, pp. 204–205).

After giving Job and Tobias as examples of endurance, he reminds his readers that

[t]his endurance the just have always had; this discipline the apostles maintained from the law of the Lord, not to murmur in adversity, but to accept bravely and patiently whatever happens in the world.... We must not murmur in adversity, beloved brethren, but must patiently and bravely bear with whatever happens, since it is written: "A contrite and humble heart God does not despise" [Ps. 50:19] (*ibid.*, 11, p. 207).

Hence, "The fear of God and faith ought to make you ready for all things", such as loss of property, diseases, loss of wife and children and other dear ones. So,

let not such things be stumbling blocks for you but battles; nor let them weaken or crush the faith of the Christian, but rather let them reveal his valor in the contest, since every injury arising from present evils should be made light of through confidence in the blessings to come. ... conflict in adversity is the trial of truth (*ibid.*, 12, p. 208).

Note Cyprian's emphasis on the activity of God and the passivity of the Christian in death. Cyprian writes that Christians who died of the current pestilence "have been freed from the world by the summons of the Lord ... *(Mortality* 20 [19], vol. 36, p. 215). Later he asserts that "those who please God are taken from here earlier and more quickly set free, lest, while they are tarrying too long in this world, they be defiled by contacts with the world". He then suggests that "when the day of our own summons comes, without hesitation but with gladness we may come to the Lord at His call." For "rescued by an earlier departure, you are being freed from ruin and shipwrecks and threatening disasters!" Therefore, "[l]et us embrace the day which assigns each of us to his dwelling, which on our being rescued from here and released from the snares of the world, restores us to paradise and the kingdom." Consider the loved ones already in heaven and the joys that await us there. "To these, beloved brethren, let us hasten with eager longing! Let us pray that it may befall us speedily to be with them, speedily to come to Christ" (26, pp. 220–221).

We see that in Cyprian's thought it is God who calls; it is he who issues the summons. God takes the Christian from the world, God frees him; God rescues him; God releases him; God restores him to heaven. The Christian is passive – he *is being* freed; he *is being* rescued; he *is being* released; he *is being* restored. This is God's activity. The Christian's activity is to yearn for heaven. Hence he should pray for an early departure from life. Yearning for death and praying to die are categorically different from taking one's own life. There is no room here for suicide. Patient endurance of all afflictions, perseverance to the end, final resignation to the will of God in the midst of those very situations that God is using to test and to refine the Christian: such thought is antithetical to the taking of one's own life.

Tertullian's message is similar:

> Let us strive, then, to bear the injuries that are inflicted by the Evil One, that the struggle to maintain our self-control may put to shame the enemy's efforts. If, however, through imprudence or even of our own free will we draw down upon ourselves some misfortune, we should submit with equal patience to that.... But if we believe some blow of misfortune is struck by God, to whom would it be better that we manifest patience than to our Lord? In fact, more than this, it befits us to rejoice at being deemed worthy of divine chastisement.... Blessed is that servant upon whose amendment the Lord insists, at whom He deigns to be angry, whom He does not deceive by omitting His admonition. From every angle, then, we are obliged to practice patience, because we meet up with our own mistakes or the wiles of the Evil One or the warnings of the Lord alike (*Patience* [19], 11.2–5, vol. 40, p. 212).

Later he gives Job as the most significant example of patient endurance:

> Far from being turned away by so many misfortunes from the reverence which he owed to God, he set for us an example and proof of how we must practice patience in the spirit as well as in the flesh, in soul as well as in the body, that we may not succumb under the loss of worldly goods, the death of our dear ones, or any bodily afflictions. What a trophy over the Devil God erected in the case of that man! What a banner of His glory He raised above His enemy.... when [Job] severely rebuked his wife who, weary by now of misfortunes, was urging him to improper remedies.... Thus did that hero who brought about a victory for his God beat back all the darts of temptation and with the breastplate and shield of patience soon after recover from God complete health of body and the possession of twice as much as he had lost (*ibid.*, 14.2–6, pp. 218–219).

Here also there is no room for suicide. The most basic principles of patient endurance for the Christian militate against the very thought of suicide. Cyprian never mentions suicide. Tertullian, however, after mentioning that Christ "tells us to give to the one who asks", remarks that, "if you take His command generally, you would be giving not only wine to a man with a fever, but also poison or a sword to one who wanted to die" (*Flight in Time of Persecution* 13.2, [19], vol. 40, p. 304). Giving wine to the febrile was thought to be very harmful. He includes assisting in suicide in the same category. The thought is that one simply will not supply the means if asked. Elsewhere he classifies anyone who "cuts his own throat" as demented or insane, and the context suggests that such a one is demon possessed (*Apology* 23.3 [19], vol. 10,. pp. 71–72). There is absolutely no suggestion in the writings of Cyprian and Tertullian that for contemporary Christians suicide either was an attraction or posed a theoretical, much less a practical, problem. Indeed all the evidence points in the opposite direction.

Of all the church fathers none was more significantly influenced by Greek philosophy than Clement of Alexandria. To him Greek philosophy was a *praeparatio evangelica* that contained more truth than falsehood. He eagerly drank from the springs of the pagan past, rejoicing in the plethora of wisdom that he felt God had revealed to the Greeks, partially through a supposed early acquaintance with Hebrew Scripture. Nevertheless, he did not accept all that these philosophers offered. How could he? Not only did they frequently contradict each other but were often at variance with Scripture. Clement's style, however, was not to draw attention to points of disagreement but, as a means of apologetics, to emphasize primarily matters of consonance. Sometimes he is rather obscure; at other times he can be easily misunderstood, especially when he draws the reader's attention to certain philosophical tenets, not to endorse them, but rather to illustrate a particular Christian truth. Here is a pertinent example:

> ... the philosophers also allow the good man an exit from life in accordance with reason, in the case of one depriving him of active exertion, so that the hope of action is no longer left him. And the judge who compels us to deny Him whom we love, I regard as showing who is and who is not the friend of God. In that case there is not left ground for even examining what one prefers – the menaces of man or the love of God (*Stromateis* 4.6 [3], vol. 2, p. 414).[41]

An "exit from life in accordance with reason" is, of course, suicide as permitted by certain philosophical schools. The Christian also has an "exit from life in accordance with reason", says Clement, and that is martyrdom. But we must remind ourselves that Clement, as we saw above, unequivocally condemns seeking martyrdom and strongly encourages flight in the face of persecution. Hence, this cannot be taken as an endorsement of suicide, as is made even more clear from Clement's discussions of suffering and sanctification that are scattered throughout his *Stromateis*.

Even when addressing suffering and sanctification, Clement's terminology is so philosophical, his vocabulary so peppered especially with Stoic jargon, that he can be easily misunderstood especially if sentences, even paragraphs, are taken out of context. The concept of *apatheia* (insensibility to suffering) is central to Stoic thought. It is also of vital importance to Clement. But in Clement it is significantly informed by those essential scriptural principles that are basic to other church fathers' values. So also for Clement suffering is an essential aspect of Christian growth and he frequently stresses God's paternal, sovereign care of his people. He says, for example, that the true Christian

will never ... have the chief end placed in life, but in being always happy and blessed, and a kingly friend of God. Although visited with ignominy and exile, and confiscation, and above all, death, he will never be wrenched from his freedom, and signal love to God. The charity which "bears all things, endures all things" is assured that Divine Providence orders all things well (*ibid.*, 4.7, p. 418).

Clement enlarges on these themes later:

...though disease, and accident, and what is most terrible of all, death, come upon [the true Christian], he remains inflexible in soul, – knowing that all such things are a necessity of creation, and that, also by the power of God, they become the medicine of salvation, benefiting by discipline those who are difficult to reform; allotted according to dessert, by Providence, which is truly good. ...he undergoes toils, and trials, and afflictions, not as those among the philosophers who are endowed with manliness, in the hope of present troubles ceasing, and of sharing again in what is pleasant, but knowledge has inspired him with the firmest persuasion of receiving the hopes of the future. [He] withstands all fear of everything terrible, not only of death, but also poverty and disease, and ignominy, and things akin to these... (*ibid.*, 7.11, p. 540–541).

Penury and disease, and such trials, are often sent for admonition, for the correction of the past, and for care for the future. Such a one prays for relief from them, in virtue of possessing the prerogative of knowledge, not out of vainglory ... having become the instrument of the goodness of God. ...he is not disturbed by anything which happens; nor does he suspect those things, which, through divine arrangement, take place for good. Nor is he ashamed to die, having a good conscience, and being fit to be seen by the Powers. Cleansed, so to speak, from all the stains of the soul, he knows right well that it will be better with him after his departure (*ibid.*, 7.13, p. 547).

The *apatheia* which Clement lauds as a Christian ideal can never logically lead to suicide:

[By] going away to the Lord ... *he does not withdraw himself from life. For that is not permitted to him.* But he has withdrawn his soul from the passions. For that is granted to him. And on the other hand he lives, having put to death his lusts, and no longer makes use of the body, *but allows it the use of necessaries, that he may not give cause for dissolution* (my emphasis, *ibid.*, 6.9, p. 497).

This is a magnificent blending of certain Stoic principles that are compatible with Clement's Christianity. But it is a Stoicism that has been Christianized to such a degree that suicide is permitted neither in the active sense of withdrawing from life nor in the passive sense of causing the dissolution of the body by failing to provide it with "necessaries".

Just as for Cyprian and Tertullian, so also for Clement, Job is the outstanding example of endurance. The Christian "will bless when under trial, like the noble Job.... He will give his testimony by night; he will testify by day; by word; by life, by conduct, he will testify" (*ibid.*, 2.20, p. 374). Later he mentions Job once more, here prefacing his praise with a commendation of the Stoics for the good qualities of their teaching:

> Fit objects for admiration are the Stoics, who say that the soul is not affected by the body, either to vice by disease, or to the virtue by health; but both these things, they say, are indifferent. And indeed Job, through exceeding continence, and excellence of faith, when from rich to become poor, from being held in honour dishonoured, from being comely unsightly, and sick from being healthy, is depicted as a good example for us, putting the Tempter to shame, blessing his Creator (*ibid.*, 4.5, p. 412).

George Fisher succinctly describes the salient and essential differences between Stoicism and early Christianity:

> Stoicism exhibits no rational grounds for the passive virtues, which are so prominent in the Stoic morals. There is no rational end of the cosmos; no grand and worthy consummation towards which the course of the world is tending. Evil is not overruled to subserve a higher good to emerge at the last. There is no inspiring future on which the eye of the sufferer can be fixed. The goal that bounds his vision is the conflagration of all things. Hence there is no basis for reconciliation to sorrow and evil. Christianity in the doctrine of the kingdom of God, furnishes the element which Stoicism lacked, and provides thus a ground for resignation under all the ills of life, and amid the confusion and wickedness of the world. For the same reason, the character of Christian resignation is different from the Stoic composure. It is submission to a wise and merciful Father, who sees the end from the beginning. Hence, there is no repression of natural emotions, as of grief in case of bereavement; but these are tempered, and prevented from overmastering the spirit, by trust in the Heavenly Father. In the room of an impassable serenity, an apathy secured by stifling natural sensibility, there is the peace which flows from filial confidence ([20], p. 175).

Fisher's assessment can be applied to Clement whose agenda was not to denigrate any philosophical school, including Stoicism, but to appropriate their best features. Although he was influenced more than any other church father by Greek philosophy (much more by Platonism than by Stoicism), nevertheless, Clement's most basic values were thoroughly Christian. Stoic *apatheia* was for him a means to godliness, a tool for sanctification through perseverance and patient endurance of all afflictions sent or permitted by the sovereign, omniscient, and paternally benevolent Deity. If any church father could have endorsed suicide under any circumstances, it would have been Clement. And in numerous places, especially in his *Stromateis*, it would have been very natural, indeed nearly inevitable, to have done so, *if* he had harbored even the remotest sympathy for it. But it was so far from his mind, so discordant with his values, and so remote from his concerns as an ethical or practical problem, that even his condemnation of it is only made in passing. This is the case also with Tertullian and, as we shall see in the next section, with the other condemnations of suicide in patristic literature before the fourth century.

DISCUSSIONS OF SUICIDE IN PATRISTIC SOURCES

Justin Martyr (ca. 100–165) imagines a pagan suggesting, "All of you, go kill yourselves and thus go immediately to God, and save us the trouble". Justin replies, "If ... we should kill ourselves, we would be the cause, as far as it is up to us, why no one would be born and be instructed in the divine doctrines, or even why the human race might cease to exist; if we do act thus, we ourselves will be opposing the will of God" (*2 Apology* 4 [19], vol. 6, p. 123). L.W. Barnard, in his insightful study of Justin, says, on the basis of this passage, that Justin "shows us men and women... who thought it a duty to preserve life so long as God delayed to take it" ([6], p. 154). There is, of course, a difference between a duty not to take one's life and a duty to preserve it, a distinction made by Justin in the context surrounding the passage quoted, in which he juxtaposes the Christian's refusal to kill himself and his willingness to die for his faith. Although this passage is a direct condemnation of suicide for Christians, it is simply an explanation to pagans why Christians do not kill themselves. Nor is it bolstered by any moral reasoning or defense. Rather, it simply maintains that it is wrong for Christians to kill themselves because God wants them in the world and the human race needs them, for without Christians there would be no one to instruct humanity in the truth. Since God sustains the human race for the sake of his people, if Christians were all removed from the world, the human race would cease to exist. His position resembles some earlier pagan and later Christian condemnations of suicide based on the premise that, since God has stationed each one in this life, no one has a right to desert that post to which God has assigned him.

The *Epistle to Diognetus*, an anonymous piece probably written in the late second century, contains a somewhat similar passage:

> The soul is locked up in the body, yet is the very thing that holds the body together; so, too, Christians are shut up in the world as in a prison, yet are the very ones that hold the world together. Immortal, the soul is lodged in a mortal tenement; so, too, Christians, though residing as strangers among corruptible things, look forward to the incorruptibility that awaits them in heaven. The soul, when stinting itself in food and drink, is the better for it; so, too, Christians, when penalized, increase daily more and more. Such is the important post to which God has assigned them, and it is not lawful for them to desert it ([37], p. 251).

There is no suggestion in either passage that suicide posed a moral problem for the Christian community. Rather the question is why Christians do not kill themselves, and the answer is that God has assigned them for an important purpose to a post that they must not abandon.

The Clementine *Homiles*, falsely attributed to Clement of Rome (late first-early second century), but written in their present form probably in the mid-fourth century, were based on an original composed in the late second or early third century. Although the *Homiles* display a marked Ebionite or Elkesaite orientation, in many ways their theology is not inconsistent with contemporary orthodoxy. In *Homily* 12 Peter encounters a pagan woman who, because of a variety of afflictions, is considering committing suicide. He admonishes her, "Do you suppose, O woman, that those who destroy themselves are freed from punishment? Are not the souls of those who thus die punished with a worse punishment in Hades for their suicide?" (12.14 [3], vol. 8, p. 295). This is a novel position, that suicide will compound one's future punishment, hence it is unfortunate that we cannot know whether this incident was in the original or added by a fourth-century redactor.

Lactantius (ca. 240–320) was a Latin rhetorician whose accomplishments attracted the attention of the emperor Diocletian, who appointed him professor of oratory in Nicomedia. He was converted to Christianity and, when the Great Persecution began (303), he felt compelled to resign his position. He turned to writing Christian apologetics directed, on the one hand, to educated pagans, and, on the other, to Christians who were troubled by philosophical attacks against their faith. The major argument of his *Divine Institutes* is that "pagan religion and philosophy are absurdly inadequate. Truth lies in God's revelation, and the ethical change which the teaching of Christ brings points conclusively to its accuracy" [42]. The first two books, "Concerning False Religion" and "Concerning the Origin of Error", attempt to refute polytheism. Book three, "Concerning False Wisdom", tries "to prove the falsity of pagan philosophy, its contradictions, and its uselessness in practice" ([26], p. 205). The remainder of the work is devoted to demonstrating the truth of Christianity. Lactantius' major statement on suicide appears in book three. Discussing the Pythagoreans and Stoics, both of whom believed in the immortality of the soul (although the latter regarded it as right to take own's own life under some circumstances), he says that many of them, "because they suspected that the soul is immortal, laid violent hands upon themselves, as though they were about to depart to heaven". He then gives as examples Cleanthes, Chrysippus, Zeno, Empedocles, Cato, and Democritus. He asserts that

nothing can be more wicked than this. For if a homicide is guilty because he is a destroyer of man, he who puts himself to death is under the same guilt, because he puts to death a man. Yea, that crime may be considered to be greater, the punishment of which belongs to God alone. For as we did not come into this life of our own accord; so, on the other hand, we can only withdraw from this habitation of the body which has been appointed for us to keep, by the command of Him who placed us in this body that we may inhabit it, until He orders us to depart from it.... All these philosophers, therefore, were homicides (3.18 [3], vol. 7, pp. 88–89).

Some years after completing his *Divine Institutes*, Lactantius was asked to produce an epitome of it. It is interesting to note that the space he devotes to suicide in this much shorter abridgement is more than in the original. In the *Epitome* he asks whether we should approve those

who, that they might be said to have despised death, died by their own hands? Zeno, Empedocles, Chrysippus, Cleanthes, Democritus, and Cato, imitating these, did not know that he who put himself to death is guilty of murder, according to the divine right and law. For it was God who placed us in this abode of flesh: it was He who gave us the temporary habitation of the body, that we should inhabit it as long as He pleased. Therefore it is to be considered impious, to wish to depart from it without the command of God. Therefore violence must not be applied to nature. He knows how to distroy His own work. And if any one shall apply impious hands to that work, and shall tear asunder the bonds of the divine workmanship, he endeavours to flee from God, whose sentence no one will be able to escape, whether alive or dead. Therefore they are accursed and impious, whom I have mentioned above, who even taught what are the befitting reasons for voluntary death; so that it was not enough of guilt that they were self-murderers, unless they instructed others also to this wickedness (39 *ibid.*, p. 237).

We should note that in the original passage, suicides are condemned as worse than homicides, for suicides desert the place to which God has appointed them. In the *Epitome* the argument is essentially the same, although he adds the offense of attempting to flee from God by committing violence against nature and the encouraging of others to do the same. His tone in the *Epitome* is even more outraged and vitriolic than in the *Institutes*. Suicides are not only homicides but are impious as well.

In passing we may observe that Lactantius' contemporary, the historian Eusebius (ca. 265 – ca. 339), writes that "tradition relates" that Pilate, the Roman official who had sentenced Jesus to death by crucifixion, "fell into such great calamity that he was forced to become his own slayer and to punish himself with his own hand, for the penalty of God, as it seems, followed hard after him" (*Ecclesiastical History* 2.7.1 [18], vol. 1, pp. 125–127). In a case such as Pilate's, suicide from despair is seen as God's penalty, a condemnation for sin, hardly setting a precedent for Christian suicide.

John Chrysostom (349–407) was a fervent and eloquent preacher whose concerns were primarily pastoral. In his *Commentary on Galatians*, when dealing with Galatians 1:4 (Jesus "gave himself for our sins to deliver us from the present evil age, according to the will of our God and Father"), he criticizes those heretics who regard the material world as evil. He takes the words 'evil world' to mean

> evil actions, and a depraved moral principle. ... Christ came not to put us to death and deliver us from the present life in that sense, but to leave us in the world, and prepare us for a worthy participation of our heavenly abode. Wherefore He saith to the Father, "And these are in the world, and I come to Thee; I pray not that Thou shouldest take them from the world, but that Thou shouldest keep them from the evil", i.e., from sin. Further, those who will not allow this, but insist that the present life is evil, should not blame those who destroy themselves; for as he who withdraws himself from evil is not blamed, but deemed worthy of a crown, so he who by a violent death, by hanging or otherwise, puts an end to his life, ought not to be condemned. Whereas God punishes such men more than murderers, and we all regard them with horror, and justly; for if it is base to destroy others, much more is it to destroy one's self ([40], vol. 13, p. 5).

Chrysostom here maintains that encouragement to suicide is a reasonable consequence of dualistic heresy. But as far as he is concerned, true Christians – "we all" would include the orthodox – regard suicides with horror, and justly so. This would be a preposterous statement, if there had been even a strong minority sentiment in the orthodox Christian community that would justify suicide to escape "this present evil world".

Ambrose (ca. 339–397), Augustine's mentor, was much influenced by both Neoplatonism and Stoicism ([17], pp. 502ff.). His position on suicide, however, seems to have been no more affected by Stoicism than that of Clement of Alexandria. In his treatise, *Death as a Good*, he comments on Paul's statement, "for to me to live is Christ and to die is gain":

For Christ is our king; therefore we cannot abandon and disregard His royal command. How many men the emperor of this earth orders to live abroad in the splendor of office or perform some function! Do they abandon their posts without the emperor's consent? Yet what a greater thing it is to please the divine than the human! Thus for the saint "to live is Christ and to die is gain". He does not flee the servitude of life like a slave, and yet like a wise man he does embrace the gain of death (3.7 [19], vol. 65, pp. 73–74).

Here once more we see the familiar assertion that one is to stay at the post to which God has assigned him until God chooses to remove him.

Elsewhere he writes his sister Marcellina, "...you make a good suggestion that I should touch upon what we ought to think of the merits of those who have cast themselves down from a height, or have drowned themselves in a river, lest they should fall in the hands of persecutors, seeing that holy Scripture forbids a Christian to lay hands on himself" (*Concerning Virgins* 3.7.32 [41], vol. 10, p. 386). Then, after giving an example of suicide by a virgin to preserve her chastity, he describes an instance of a woman's endurance under torture leading to death, implying, in answer to the question raised above, that suicide undertaken to avoid persecution is wrong but to preserve virginity is laudable. It is noteworthy that Ambrose simply asserts that Scripture forbids suicide and appears to assume that his addressee would share this opinion. This, incidentally, is apparently the earliest blanket appeal to Scripture for a condemnation of suicide.

Very similar are the views of Jerome (ca. 345 – ca. 419). In a letter to which reference has already been made above, written to the lady Paula, who was distraught over the death of her daughter Blaesilla, he says,

Have you no fear, then lest the Savior may say to you: "Are you angry, Paula, that your daughter has become my daughter? Are you vexed at my decree, and do you, with rebellious tears, grudge me the possession of Blaesilla? You ought to know what my purpose is both for you and for yours. You deny yourself food, not to fast but to gratify your grief, and such abstinence is displeasing to me. Such fasts are my enemies. I receive no soul which forsakes the body against my will. A foolish philosophy may boast of martyrs of this kinds; it may boast of a Zeno, a Cleombrotus, or a Cato. My spirit rests only upon him 'that is poor and of a contrite spirit and that trembleth at my word; Is. 66:2'" (*Letter* 39.3 [35], vol. 6, p. 51).

Elsewhere, Jerome qualifies this otherwise unlimited condemnation of suicide: "It is not ours to lay hold of death; but we freely accept it when it is inflicted by others. Hence, even in persecutions it is not right for us to die by our own hands, except when chastity is threatened, but to submit our necks to the one who threatens" (*Commentarius in Ionam Prophetam* 1.6 [35], 1.12.390–391).

We should note that both Ambrose and Jerome make one exception to their condemnation of suicide: when committed to preserve one's chastity. However, they do condemn suicide to escape persecution. The latter was pretty much academic for the orthodox community by this time, since persecution leading to martyrdom had generally ceased to be a threat with the legalization of Christianity in 313. The only specific example of any Christians committing suicide to avoid martyrdom is reported by Eusebius in his *Ecclesiastical History*, where he expresses neither approval nor disapproval:

Why need one rekindle the memory of those in Antioch, who were roasted on heated gridirons, not unto death, but with a view to lengthy torture; and of others who put their right hand into the very fire sooner than touch the accursed sacrifice? Some of them, to escape such trials, before they were caught and fell into the hands of those that plotted against them, threw themselves down from the tops of lofty houses, regarding death as a prize snatched from the wickedness of evil men (8.12.2 [18], vol. 2, p. 298).

This passage ought to be interpreted in light of the implied condemnation of this type of suicide by Ambrose (*Concerning Virgins* 7.32–39) and the very specific condemnation of it by Jerome (*Commentarius in Ionam Prophetam* 1.6).

Both Ambrose and Jerome assume, however, the probity of suicide to preserve chastity. Here we enter into an anomaly in early Christian thought: the approbation of suicide to preserve chastity, especially to preserve virginity. Only a minority of the sources before Augustine mention it; the few that do, approve it.[42] The earliest of these is Eusebius. He tells of a virtuous woman and her two virgin daughters at Antioch who threw themselves into a river to escape the salacious designs of the Roman soldiers (*Ecclesiastical History* 8.12.3–4). Later, after describing the endurance of martyrs under the most severe tortures, he writes:

And the women, on the other hand, showed themselves no less manly than the men, inspired by the teaching of the divine Word: some, undergoing the same contests as the men, won equal rewards for their valour; and others, when they were being dragged away to dishonour, yielded up their souls to death rather than their bodies to seduction (=*phthora*= 'moral corruption'; 8.14.14, [18], vol. 2, p. 309).

Then he gives an example of a "most noble and chaste" married lady in Rome who, when faced with imminent threat of sexual violation,

transfixed herself with a sword. And straightway dying she left her corpse to her procurers; but by deeds that themselves were more eloquent than any words she made it known to all men, both those present and those to come hereafter, that a Christian's virtue is the only possession that cannot be conquered or destroyed (*ibid.*, 8.14.17, p. 311).

Eusebius' approval of these cases is obvious.

We have already noted that Ambrose appears to approve such suicide. The remarks of the leading twentieth-century authority on Ambrose, F. Homes Dudden, are worthy of inclusion here:

On one occasion Ambrose was requested by his sister Marcellina to state his opinion concerning virgins who committed suicide to avoid violation. He did not, however, express himself very clearly. On the one hand, he told the story of the suicide of St. Pelagia of Antioch and of her mother and sisters in a manner which suggests that, if he did not actually commend, he certainly did not condemn their act. On the other hand, he spoke with unqualified admiration of another Antiochene virgin, who, being sentenced to violation in a brothel, on account of her refusal to sacrifice, prepared to undergo the penalty, without anticipating it by suicide. In relating the incident, he said, "A virgin of Christ may be dishonoured, but she cannot be polluted. Everywhere she is the virgin of God and the temple of God. Places of infamy do not stain chastity; on the contrary chastity abolishes the infamy of the place" ([17], p. 157).

Ambrose's reasoning, as expressed in the last three sentences quoted here from his treatise *Concerning Virgins*, corresponds very closely to Augustine's, who, as we shall see, on these very grounds condemns suicide even to preserve chastity.

Jerome, as has already been observed, is unequivocal in his approbation of these acts of suicide. For him, they are the only exception to his firm conviction that suicide is illicit for Christians. In order to appreciate why such an exception was made by these church fathers, one must be aware of the extent to which the early Christian community recoiled from what they regarded as the gross immorality of pagan culture. Early Christian sources condemned nearly all aspects of pagan immorality as related features of a society which they regarded as rotten to its very core because of sin. With sweeping strokes of moral indignation, early Christian apologists condemned gladiatorial shows and cruel executions, along with abortion, infanticide, homicide generally, and a broad variety of sexual practices ranging from homosexuality to adultery, from fornication to perversions within the marriage bond. The range of imagination employed in sexual activities by the pagans, and the open display of it, appears to have caused the early Christian community to react more strongly against sexual sins than against any other realm of contemporary immorality. And the numerous injunctions in the New Testament to refrain from sexual sins supported their moral indignation.

But these factors alone were not sufficient to produce a climate conducive to regarding the preserving of one's chastity as a higher moral obligation than refraining from suicide. Sexual abstinence is occasionally praised or recommended in the New Testament, and as early as the late first or early second century Ignatius calls Christian women who voluntarily remain virgins Christ's brides and jewels. But it is not until the mid- to late second century that some sources begin to suggest that celibacy is a higher good than marriage and an essential quality for the true ascetic. That virginal celibacy was continuing to grow in esteem, gradually becoming nearly the highest virtue, is illustrated by the fact that instances of Christian women committing suicide to preserve their chastity are not found before the fourth century. It is this one (and by then approved) motivation for suicide that first caught Augustine's attention and caused him to address the subject of suicide.

All the implicit and explicit condemnations of suicide in Christian literature prior to or contemporary with Augustine are encapsulated and elaborated by Augustine. There is no figure in the early centuries of Christianity who had a more significant influence on later western Christian thought than he. So similar was he to his eastern contemporaries, Basil of Caesarea, Gregory Nazianzen, and Gregory of Nyssa (the Cappadocian Fathers), and to John Chrysostom that to see him as a spokesman for western (as opposed to eastern) Christianity would be simplistic. Yet there is in Augustine's works a melody that is hauntingly 'medieval' in many of its nuances. The end of his long life marks a watershed in western history. When he died in 430, Rome had already been sacked (twenty years earlier) by the Goths and Roman North Africa was falling to the Vandals. The unity of the Mediterranean world was beginning to disintegrate as its western half was increasingly affected by forces from the north, some already Christianized, and was slowly being drawn into an emerging European, rather than a strictly Mediterranean, ethos.

Hence we need to approach Augustine as something of a Janus figure. His influence on medieval Christianity is enormous, and we shall conclude the present study with some brief comments on his significance for medieval treatments of suicide, about which there is no pressing controversy. Much more important is a consideration of the extent to which he resembles and transmits the values of his antecedents and contemporaries as they pertain to a consideration of suicide. As early as Rousseau, some have claimed that Augustine took his arguments against suicide from Plato's *Phaedo* and not from the Bible. It is common to find modern scholars making such amazing claims as "there is little reason to think that Augustine's position is authentically Christian..." ([8], p. 71). It should be obvious to the attentive reader that a survey of the patristic literature demonstrates that it is simply wrong to suggest that Augustine formulated what then became the 'Christian position' on suicide. Rather, by removing certain ambiguities, he clarified and provided a theologically cogent explanation of and justification for the position typically held by earlier and contemporary Christian sources.

Augustine's best-known discussion of suicide is found in Book One of *City of God*. This is a digression that could not have been intended to be a systematic and comprehensive treatment of the moral issues involved in a consideration of suicide, regardless of the extent to which portions of the latter were used as authoritative by later generations. His discussion of suicide must be appreciated within the context of his immediate purposes in formulating a position to which a consideration of suicide is only incidental.

In his introductory letter to the first installment of this work, published in 414, Augustine says that he had undertaken the writing of the work in order to defend the City of God, that is, the community of those "predestined to reign with God from all eternity" (as he eventually defines it, 15.1 [5], p. 595) against the pagans. Just four years earlier (410), Rome had been captured and ravaged by Alaric and Goths. This had sent a shockwave throughout the Empire. Even though pagan temples had been closed by imperial degree about two decades earlier and public worship of the traditional deities forbidden, the city of Rome had remained largely pagan, especially its upper classes, who clung tenaciously and defiantly to their ancient religious practices, refined by the increasingly popular Neoplatonism that had become nearly a religion itself. The sack of Rome was, in the pagans' opinion, the final proof of the gods' displeasure with the official neglect of their worship. It was in response to such sentiments that Augustine began, essentially as an encouragement to his fellow Christians in the face of troubling accusations by pagans, what was, over the next thirteen years, to evolve into the *City of God*.

Shortly after finishing this massive work, Augustine wrote his *Retractations* in which he describes the schema of the *City of God*. Here he says that the first five books were designed to refute those who tie the prosperity of the Empire to the favor of the gods and adversity to their disfavor (*Retractations* 2.69). Book One begins with the assertion that Christianity had mitigated rather than aggravated the violence of the recent sack of Rome in particular and of war in general. Christians show a clemency antithetical to the typical savagery of the pagans (1.1–7). He argues that prosperity and adversity affect both the good and the bad alike (1.8–9). Christians lose nothing when deprived of temporal goods, even of life itself, since all must die sooner or later (1.10–11). Furthermore, it is not a matter of great concern whether the dead bodies of Christians are abused and left unburied, as happened in several instances during the recent sack of Rome. Nevertheless, if possible, Christians pay respect to the bodies of their dead (1.12–13). He then describes the consolations that Christians experience when in captivity and reminds the pagans of their own Regulus, who, centuries before, had endured captivity in an exemplary manner for the sake of his religion, albeit false (1.14–15). Some pagans took great delight in pointing out that even Christian women were sexually violated by the barbarian Goths during their ravaging of Rome. Augustine replies that the virtue of one thus violated is not polluted (1.16). It is here that Augustine detours into a discussion of suicide by women to preserve their chastity.[43]

Who is so lacking in human compassion, he asks, as to refuse to excuse them for doing this? But other women did not kill themselves under the same circumstances, because they did not want to escape "another's criminal act by a crime of their own". Anyone who faults the latter lays himself open to a charge of folly. At this point he turns from a consideration of this specific type of suicide to make his *first general condemnation of suicide*, which rests on two grounds:

1. If no one on his own authority has a right to kill even a guilty man, then certainly one who kills himself is a homicide. The more innocent he is in respect to that for which he puts himself to death, the more guilty he is for killing himself.
2. We rightly execrate Judas' deed, and truth declares that when he hanged himself he increased rather than atoned for his detestable betrayal. Truth declares this to be so because by killing himself, Judas, in despairing of God's mercy while displaying a self-destructive sorrow, left no room for a saving repentance. How much more, then, ought one who has no fault in himself worthy of such a punishment refrain from self-slaughter. When Judas killed himself, he killed a criminal. He died guilty not only of Christ's death, but of his own as well.

Augustine returns to the initial question: what of those who commit suicide to preserve their chastity? "Why then should a person who has done no evil do evil to himself, thus killing an innocent person lest he have to submit to another's wrongdoing and in so doing perpetrate his own sin upon himself, lest another's sin be perpetrated upon him?" Here ends 1.17.

Augustine devotes 1.18 to the question whether fear of being morally polluted by another's lust is legitimate. Augustine's answer is a resounding "no". If the lust is another's there is no moral pollution for the one victimized, since purity is a virtue of the mind. We need not consider his arguments here. His conclusion to this section is that a woman who has been violated by another's sin has in herself no fault worthy of being punished by voluntary death. How much less, then, is she right in killing herself to ensure that she not be violated. He then expresses a strong admonition: Let there be no certain murder when the offense itself, although another's, still is only pending and uncertain.

Having established that guilt attaches only to the ravisher if the will of the victim does not consent to the act, he considers in 1.19 the case of the legendary Roman matron Lucretia, who had killed herself because she had been violated and was unable to endure living with the horror of the indignity that she had suffered. The Romans venerated her as a paragon of virtue, but, says Augustine, in killing herself she received the greater punishment, since the ravisher was only exiled. What kind of justice is this? And, after all, her suicide was not due to any great value that she placed on chastity but, rather, was due to the weakness of shame. Her sense of honor could not tolerate the thought that some might think that she had willingly submitted to an adulterous sexual act. But this is not the way Christian women acted who were violated but still survived. Not only did they not avenge another's crime, but they would not compound the wrong by adding crimes of their own by committing murders against themselves. They have within themselves the glory of chastity, the witness of their conscience. This they have in the sight of God and they do not ask for anything more. There is, indeed, nothing more for them to do that they could rightly do, "lest they deviate from the authority of divine law while doing wrong to avoid people's suspicion."

In 1.20 Augustine makes his *second general condemnation of suicide*, on the following grounds:

1. It is significant, he says, that in the sacred canonical books no divinely given command or permission can be found for us to kill ourselves either to attain immortality or to avoid or escape any evil.
2. Quite to the contrary, the killing of oneself must be understood to be forbidden when the law says, "You shall not kill", especially because it does not add "your neighbor", as it does when it forbids bearing false witness. Even though the commandment against bearing false witness against one's neighbor also should be understood to include the prohibition against bearing false witness against oneself, the absence of

the addition of "your neighbor" to the commandment not to kill shows that there is no exception, not even the very one to whom the command is addressed.

Now he maintains in 1.21 that not all killings of men are acts of homicide, e.g., in obeying legitimate orders in executing criminals or in war. He then gives some extraordinary examples of legitimate killings. Abraham was prepared to kill Isaac at God's command and would rightly have done so if God had not intervened. Samsom destroyed himself along with his enemies. God's Spirit, who had previously worked miracles through him, must certainly have ordered him to do this. He concludes this section by saying that with these two exceptions, (i.e., killing prescribed by a just law or directly by God), whoever kills a human being, either himself or anyone at all, is entwined in the crime of homicide.

He next poses the question whether suicide is ever a sign of greatness of soul. Those who have killed themselves are perhaps to be admired for their greatness of soul but not praised for the soundness of their wisdom. But if one considers the matter carefully, greatness of soul may not properly be applied to one who has killed himself because he lacked the strength to endure hardship or another's wrongs. An inability to bear physical oppression or the stupid opinion of the rabble is rather the sign of weakness of character. If suicide can be taken as a sign of greatness of soul, then Theombrotus should be an outstanding example: When he read Plato's dialogue that discusses the immortality of the soul, he went right out and killed himself so that he could go immediately to a better life. And Theombrotus had not been suffering under any hardship whatsoever. Would not Plato himself have done the same thing if that intellect by which he perceived the soul's immortality had not also shown him that such an act must be forbidden?

Many, of course, have killed themselves rather than fall into the hands of the enemy. But the patriarchs, prophets, and apostles did not do so. Indeed, Christ advised the latter that when they were persecuted in one town they should flee to another. He could have advised them to kill themselves to avoid falling into the hands of their persecutors. Instead, he promised to prepare eternal mansions for them. Augustine ends 1.22 by asserting that, regardless of what kinds of examples many pagans give of their ancestors thus ending their own lives, "it is obvious that this is not right for those who worship the true God".[44]

In the next two sections he contrasts two examples from Rome's past. Cato, whom the Romans regarded as a man of learning and integrity, committed suicide to avoid falling into Caesar's hands. Why did he commit suicide? Augustine suggests that he apparently did not want to allow Caesar to receive the praise he would win by pardoning him. Augustine begins 1.24 by asserting that Job, who chose to continue to suffer horrible physical distress rather than kill himself, is to be preferred over Cato. So also are other saints mentioned in the sacred writings who chose to endure captivity and slavery rather than to end their own lives. Now he brings up a virtuous pagan to contrast with Cato; that is Regulus, whom Augustine had already given as an example of patience endurance (1.15). Since Regulus chose to stay alive under the most trying of circumstances, there can be no doubt that he judged it to be a great crime for a person to kill himself. Augustine says that he could give other examples of pagans who had no fear of death but yet chose to endure domination by the enemy rather than to inflict death on themselves. "How much more will Christians, who worship the true God and long for a heavenly country, abstain from this crime". For Christians assume, under such circumstances, that Divine Providence has subjected them to enemies either to try or to correct them. For God does not anbandon them in their state of humiliation. Furthermore, since they are under no obligation to kill a conquered enemy (*hostis*), "what, then, is the source of this evil error that a person should kill himself, either because his enemy (*inimicus*) has sinned against him or may sin against him, although he would not dare to kill even an enemy (*inimicus*) who had sinned or may sin against him?" With this question Augustine ends 1.24 and is obviously working his way back to the question that precipitated his digression on suicide.

In 1.25 he deals with the proposition that one sin should not be avoided by the commission of another sin. There is always the possibility that when subjected to another's lust one may be enticed to consent to the sin. Accordingly, some say that one so threatened ought to kill himself, on the ground not of another's sin, but rather on the ground of the potential for one's own sin. But, replies Augustine, a mind that is subject to God and his wisdom rather than to the body and its lusts will never consent to the physical desire aroused by another's lust. "In any event, if the killing of oneself is a detestable crime and a damnable sin just as obvious truth proclaims, who is so stupid as to say, 'Let us now sin immediately, lest perhaps we sin later; let us now commit homicide immediately lest perhaps we fall into adultery later'?" Surely it is better to take a chance on an uncertain adultery in the future than on a certain homicide now. Is it not better to commit a crime that repentance may heal than an act of wickedness that affords no opportunity for repentance? Augustine says that he includes these comments "for the sake of those men or women who think that they must do mortal violence to themselves in order to avoid, not another's sin, but a sin of their own." He proceeds to assure them that the mind of a Christian who trusts in his own God, hoping in him and relying on his help, will not consent to participation in another's sin. He will resume discussion of the subject in 1.27, the final section of this digression on suicide.

Augustine has clearly condemned as sinful, but understandable, the suicides of Christian women who considered death preferable to being ravished by barbarians during the recent sack of Rome. But what of those women of the past who, in the face of persecution, had chosen to commit suicide to preserve their chastity and have not only become heroines of the faith but also are venerated as martyrs in the Catholic Church by throngs of people who visit their shrines? It is with this difficult question that he introduces 1.26, a section that addresses the broader question, "What explanation should be given for those unlawful acts committed by saints?" Regarding those women who killed themselves during times of persecution in the past and are now venerated as martyrs, he does not "dare to give a rash judgment". Perhaps divine authority instructed the church by some trustworthy evidence that their memory should be honored. Maybe these women acted not under human misconception but by divine command. In this case, they were not erring but obeying. But this he also has no way of knowing. If the latter is true, then the case is comparable to Samson's. There must be absolutely no doubt about the certainty of the divine command. He concludes this section with the assertion that

> ...no one should inflict a voluntary death on himself by fleeing temporal troubles, lest he fall into eternal troubles; that no one ought to do so on account of another's sins, lest by this very act he create his own very serious sin, when he would not have been polluted by another's sin; that no one ought to do so because of past sins, because he needs his present life all the more so that his sins may be healed by repentance; that no one ought to do so from a desire for the better life for which we hope after death, because this better life, which comes after death, does not receive those responsible for their own death.

This could have been a quite appropriate concluding statement of this digression, as it deals with a variety of hypothetical exceptions to the prohibition of suicide. It is noteworthy that the hypothetical exception not included here is the one to which the immediately preceding section was devoted, to which Augustine feels he must devote more attention.

In 1.27 Augustine resumes the question whether one may commit suicide to escape being lured into sin by enticing pleasure or impelled into sin by raging pain. If we ever consent to this there is no drawing the line: then why not kill oneself immediately after being baptized? For what reason would anyone chose to endure the pressures of life with its temptations? Why waste our time exhorting to holiness and to the avoiding of sin when we could persuade people to take a shortcut that would avoid all risks of sin? That would not be foolishness but madness. Because it is wicked even to suggest this, it is certainly wicked to kill oneself. "For if any just cause were possible for suicide, I am sure that there could not be one more just than that. But since not even this one is just, therefore there is no just cause for suicide". Here ends the digression on suicide, and Augustine resumes the subject that had precipitated this disgression with a word of encouragement: "So, faithful Christians, don't regard your life as a burden because your enemies mock your chastity" (1.28).

Augustine has covered the bases very well in condemning different motivations for suicide: 1) a desire to escape from or avoid temporal troubles; 2) a desire to escape from or avoid another's sinful actions (suicide to protect one's chastity would be included here); 3) guilt over past sins; 4) a desire for the better (i.e., eternal) life; and 5) a wish to avoid sinning. He emphatically maintains that if there were any just cause for suicide, it would be the last. He makes only passing mention of martyrdom in this digression on suicide, and that is in his refutation of the pagans' approval of suicide to avoid captivity. He says that the partiarchs, prophets, and apostles certainly did not do so, and in reference to the apostles quotes Jesus' admonition that when persecuted in one city, they should flee to another (1.22). He will shortly be devoting much attention to the subject, but it will not be to the martyrdom of fellow Catholics; rather, it will be to the courting of martyrdom by, and the 'heroic' suicides of, members of a schismatic/heretical group, the Donatists.[45]

The Donatist movement was named after its leader, Donatus, who was bishop in Carthage from 313 to about 355. The movement had actually been formed two years earlier by rigorists who condemned what they regarded as the laxness of the church in accepting back into fellowship those who had apostatized during the Great Persecution under Diocletian. It was essentially a protest movement that regarded itself as the one true and holy church. The Catholics, they maintained, because of their toleration of low standards of holiness, were apostates. The Donatists saw themselves as upholders of purity of discipline in the face of Catholic compromise with the 'world' and its system, including the ostensibly Christian emperors. From its inception, the sect was both on the offensive and on the defensive; they were the church of the righteous who alone were pure.

The schismatic basic for the Donatist movement evoked a negative response both from the Catholic Church and from the Emperor Constantine, who, beginning in 317, attempted to coerce them back into the fold. Persecution encouraged rather than discouraged these rigorists and merely confirmed them in their convictions. They appear to have been only spasmodically persecuted until early in the fifth century when a series of repressive measures were promulgated by imperial edict. In 415 the death penalty was specified for those Donatists who continued to assemble.

Augustine greeted these draconian measures with enthusiasm, except for the death penalty, which he consistently opposed in principle: it, like suicide, removed any possibility of repentance (*Epistle* 153.18). Furthermore, it provided the Donatists with the martyrdom that so many of them seem to have wanted. Augustine's position on coercion had changed over the years along with his evolving ecclesiology. Earlier he had maintained that only spiritual measures should be used against heretics. Now he advocated force to bring them into the Catholic Church, for outside the Catholic Church there could be no salvation. He applied the *cogite intrare* of Luke 14:23 to the treatment of heretics by civil authorities, hoping that by these measures they could be saved when reason and instruction failed.

Undoubtedly a catalyst to Augustine's change of heart on the question of religious coercion were those aspects of the Donatist movement that he found exceptionally objectionable: their provoking persecution through acts of violence and obnoxious defiance, and their 'heroic' suicides. The most noteworthy practitioners of these tactics that Augustine so loathed were the Circumcellions, a fringe group of Donatists who were generally an embarrassment to the more moderate members of the sect. The Circumcellions often roamed the countryside, inciting peasants and slaves to rebellion, engaging in indiscriminate as well as systematic acts of violence, destroying Catholic churches, harassing, sometimes maiming, and occasionally even killing Catholic clergy. They once attempted to kill Augustine by an ambush that he escaped only by having accidentally taken the wrong road on a journey (Possidius, *Vita Augustini* 12). It is impossible to determine with accuracy when it is fair to saddle the entire Donatist movement with complicity in the actions of the Circumcellions, or to know when the fanatical actions of the latter were simply an extension of the ideology of the former, especially in their provoking of martyrdom and in their 'heroic' suicides.

Augustine, perhaps unfairly, attributes both of these acts quite indiscriminately to the Donatists generally. In one of his earliest anti-Donatist works, the *Contra litteras Petiliani*, written between 401 and 405, he asks, "[I]f you are suffering persecution, why do you not retire from the cities in which you are, that you may fulfill the instructions" of Christ to flee when persecuted? Why are you always "eager to annoy the Catholic Churches by the most violent disturbances, whenever it is in your power, as is proved by innumerable instances...?" (2.19.43 [40], vol. 4, pp. 539–540). But what disturbed him far more was their practice of 'heroic' and very dramatic suicides. Just before committing suicide, they would shout "Deo laudes": "You are so furious, that you cause more terror than a war trumpet with your cry of 'Praise to God'; so full of calumny, that even when you throw yourselves over precipices without any provocation, you impute it to our persecutions" (2.85.186, *ibid.*, p. 574). This was written before the death penalty had been imposed against Donatists who persisted in assembling, and apparently Augustine's point is that their propensity to suicide was so ingrained that their acts of self-destruction could not be mitigated by the claim that it was done to avoid being put to death by their persecutors (*ibid.*, 2.20.46, p. 541).

A few years later (416), after the promulgation of several decrees against the Donatists, including the death penalty for those who continued to assemble, Augustine wrote to Donatus, a priest of the Donatist sect, who had been arrested. Augustine writes:

> You are angry because you are being drawn to salvation, although you have drawn so many of our fellow Christians to destruction. For what did we order beyond this, that you should be arrested, brought before the authorities, and guarded, in order to prevent you from perishing? As to your having sustained bodily injury, you have yourself to blame for this, as you would not use the horse which was immediately brought to you, and then dashed yourself violently to the ground; for, as you well know, your companion, who was brought along with you, arrived uninjured, not having done any harm to himself as you did (*Letter* 173.1 [40], vol. 1, p. 544).

Augustine then argues that it is "fitting that you should be drawn forcibly away from a pernicious error, in which you are enemies to your own souls..." (173.2, *ibid.*). The "pernicious error" to which he refers is two-fold: in the broadest sense it is the heretical teaching of the Donatist sect that would lead to spiritual death; in a narrower sense it is the act of suicide, from which Donatus was forcibly restrained, which would have caused physical as well as spiritual death. Augustine maintains that if

> mere bodily safety behooves to be so guarded that it is the duty of those who love their neighbour to preserve him even against his own will from harm, how much more is this duty binding in regard to that spiritual health in the loss of which the consequence to be dreaded is eternal death! At the same time let me remark, that in that death which you wished to bring upon yourself you would have died not for time only

but for eternity, because even though force had been used to compel you – not to accept salvation, not to enter into the peace of the Church, the unity of Christ's body, the holy indivisible charity, but - to suffer some evil things, it would not have been lawful for you to take away your own life (*ibid.*, 173.3 and 4, p. 545).

He next argues that the Scriptures give no precedents for such self-destruction and seeks to refute the Donatist claim that 1 Cor. 13:3 does:

I have heard that you say that the Apostle Paul intimated the lawfulness of suicide, when he said, "Though I give my body to be burned", supposing that because he was there enumerating all the good things which are of no avail without charity, such as the tongues of men and of angels, and all mysteries, and all knowledge, and all prophecy, and the distribution of one's goods to the poor, he intended to include among these good things the act of bringing death upon oneself. But observe carefully and learn in what sense Scripture says that any man may give his body to be burned. Certainly not that any man may throw himself into the fire when he is harassed by a pursuing enemy, but that, when he is compelled to choose between doing wrong and suffering wrong, he should refuse to do wrong rather than to suffer wrong, and so give his body into the power of the executioner, as those three men did who are being compelled to worship the golden image, while he who was compelling them threatened them with the burning fiery furnace if they did not obey. They refused to worship the image: they did not cast themselves into the fire, and yet of them it is written that they "yielded their bodies, that they might not serve nor worship any god except their own God". This is the sense in which the apostle said, "If I give my body to be burned" (173.5, *ibid.*).

Augustine then argues that the Donatists' interpretation of 1 Cor. 13:3 is incorrect since they are devoid of charity both in their actions and in their state and ends with an appeal to the *cogite intrare* of Luke 14:23 as justification for employing coercion against recalcitrant Donatists.

A year later Augustine wrote a treatise entitled *Concerning the Correction of the Donatists* in the form of a letter (number 185). The Donatists can be easily distinguished from true martyrs, Augustine maintains, for

true martyrs are such as those of whom the Lord says, "Blessed are they which are persecuted for righteousness sake". It is not, therefore, those who suffer persecution for their unrighteousness, and for the divisions which they impiously introduce into Christian unity, but those who suffer for righteousness' sake, that are truly martyrs (185.2.9 [40], vol. 4, p. 636).

Some, however, may think that the Donatists have been driven to suicide by persecution. That is not so, says Augustine, because when they were not being persecuted and the pagan temples were still permitted to be open, they would come to the temples and provoke the pagans to kill them. Others

went so far as to offer themselves for slaughter to any travellers whom they met with arms, using violent threats that they would murder them if they failed to meet with death at their hands. Sometimes, too, they extorted with violence from any passing judge that they should be put to death by executioners, or by the officer of his court. ...it was their daily sport to kill themselves by throwing themselves over precipices, or into the water, or into the fire. For the devil taught them these three modes of suicide, so that, when they wished to die, and could not find any one whom they could terrify into slaying them with the sword, they threw themselves over the rocks, or committed themselves to the fire or the eddying pool (*ibid.*, 185.3.12, pp. 637–638).

That the devil is the cause of the Donatists' acts of self-destruction by no means exculpates them. Rather, it is for Augustine evidence of the extent to which these acts of 'heroic' suicide are profoundly execrable.

In a letter written in 420, Augustine makes the interesting argument that he who thinks that it is "advantageous and allowable" to kill himself should also kill his neighbor,

since the Scripture says: "Thou shalt love thy neighbour as thyself". But when no laws or lawful authorities give comment, it is not lawful to kill another, even if he wishes and asks for it and has no longer the strength to live, as is clearly proved by the Scripture in the Book of Kings, where King David ordered the slayer of King Saul to be put to death, although he said that he had been importuned by the wounded and half-dead king to kill him with one blow and to free his soul struggling with the fetters of the body and longing to be released from those torments. Therefore, since everyone who kills a man without any authorization of lawful power is a murderer, anyone who kills himself will not be a murderer if he is not a man. I have said all this in many ways in many other sermons and letters of mine (204 [19], vol. 32, p. 6).

This point, however, that it is a sin to aid in the suicide of one who strongly desires to die but does not have the strength to kill himself, does not appear elsewhere in any of his discussions of suicide.

Now Augustine brings up a matter which he says that he does not recall ever having addressed before. The Donatists, "embarrassed by the extreme scarcity of examples" that would set "a precedent for the crime of self-destruction", claim that they have found one in Maccabees (2 Mac. 14:37–46), specifically in the person of Razias who had taken his own life rather than submit to captivity by the enemy. Augustine says of him:

Since he was in high esteem among his own, and was most zealous in the Jewish religion...and since for this reason this same Razias was called the father of the Jews, what wonder if an overweening pride found its way, so to speak, into the man so that he chose to die by his own hand rather than suffer the indignity of slavery at the hands of an enemy after having enjoyed such eminence in the sight of his countrymen? Deeds like that are usually praised in pagan literature. But although the man himself is praised in these Books of the Maccabees, his deed is merely related, not praised, and it is set before our eyes as something to be judged rather than imitated.... Obviously, Razias was far from those words which we read: "Take all that shall be brought upon thee, and in thy sorrow endure, and in thy humiliation have patience", [Ecclus. 2:4]. Therefore he was not a man of wisdom for choosing death, but of impatience in not bearing humiliation (*ibid.*).

Patience – that is, patient endurance – was just as fundamental a principle for Augustine as it was for earlier and contemporary church fathers. Patient endurance was to Augustine a distinct quality of God's people, whether of the Old Covenant or of the New. In 415 he had written a treatise entitled *De Patientia*, which begins with the statement that patience is a virtue of the soul that is not only a gift of God but is predicated of God himself. This treatise is similar in much of its emphasis to earlier works on patience that we have already considered. The major difference is that the earlier discussions, written as they were while Christianity was an illicit cult, devoted much attention to patient endurance of persecution. Like earlier authors, Augustine presents Job as the supreme example of patience:

At him let those men look who bring death upon themselves when they are being sought out to be given life, and who, by taking away their present life, reject also the life to come. For, if they were being forced to deny Christ or to do anything contrary to justice, they ought, as true martyrs, to bear all things patiently rather than to inflict death upon themselves in their impatience. If he could have done it righteously to escape evil, holy Job would have destroyed himself so that he might have escaped such diabolic cruelty in his own possessions, in his own sons, in his own limbs. But he did not do it. Far be it that a wise man commit against himself what not even his foolish wife suggested. Because, if she had suggested it, she would deservedly have had the reply which she heard on suggesting blasphemy: "Thou hast spoken like one of the foolish women: if we have received good things at the hand of God, why should we not receive evil?" And, had he lost his patience either by blaspheming, as she had wished, or by killing himself, which she had not dared to suggest, he would have died and would be among those about whom it has been said: "Woe to them that have lost patience". And he would have increased rather than escaped punishment, for, after the death of his body, he would be hurried away to the penalties of blasphemers or homicides or the more grievous ones of parricides. For, if parricide is more heinous than any homicide in that one slays not merely a man, but one's neighbor, and in that type of murder one's guilt is more serious the closer the person one has destroyed, then without doubt he is a worse sinner who commits suicide, for no one is closer to a man than himself. What now are those wretched men doing, who suffer self-inflicted punishments here and afterwards pay the penalty due, not only for their impiety toward God but also for their cruelty toward themselves? And then they look for the glory of martyrdom! Even if they were suffering persecution in order to bear witness to Christ, and killed themselves so as not to suffer anything from their persecutors, it would rightly be said of them: "Woe to those who have lost patience". For, how could the reward of patience be given to them justly if it was impatient suffering that was to be crowned? Or if he murders himself, a crime which he is forbidden to commit against a neighbor, how will he, to whom it has been said: "Thou shalt love thy neighbor as thyself", be judged innocent? (*Patience*, 13.10, [19], vol. 16, pp. 246–248).

It is interesting to note that Augustine now introduces the argument that the degree of heinousness of murder depends on the degree of propinquity of the murdered to the murderer. Hence suicide is the most reprehensible type of murder. More significant overall, however, is that suicide, for Augustine, is the fruit of the lack of patient endurance. At the beginning of this treatise, he had told his readers that patience is a gift of God, and the treatise ends on that extended note. Patience will also prove to be the climax of Augustine's second digression on suicide in the *City of God*, which is his final statement on the subject.

In 426 or 427 Augustine published Book Nineteen of the *City of God* along with the remainder of the work. This, the last section of Augustine's *magnum opus*, opens with a discussion of three pagan views of life's *summum bonum* and *summum malum*. Augustine then presents a fourth that is opposed to these pagan perspectives, all of which maintain that the *summum bonum* for oneself resides in oneself. For the Christian, the *summum bonum* is eternal life and the *summum malum* is eternal death. 'Truth' laughs at those who place the *summum bonum* in such things as the body or soul, pleasure or virtue, or in anything that they think they can achieve by their own efforts. What pain and turbulence may not strike the wise man's body? What if the mind is affected by the senses, e.g., if one becomes deaf or blind, or if one is rendered insane by some disease? The insane say or do many senseless things, mostly alien or even opposed to their own purpose and character. And what of those who suffer the assaults of demons? Where is their intellect when the evil spirit is using their souls and bodies according to its own will? And who is confident that this evil cannot happen to the wise man in his life? As to the virtues (temperance, prudence, justice, fortitude, and patience), there can be nothing but perpetual warfare with internal vices.

How, then, Augustine asks, can the Stoics claim that all these ills that beset the body, the mind, and the will are not ills at all, since they admit that if the wise man cannot or ought not endure them, he is compelled to inflict death on himself and depart from this life? So great is their stupidity that they call their wise man 'happy', even if he becomes blind and deaf and dumb, loses the use of his limbs, is racked with pain and afflicted with every other imaginable evil, and finally is driven by these things to kill himself. What a 'happy' life it is that seeks to end itself! If it is 'happy', let him remain in it. But these things must be evil that conquer that good which is called fortitude and compel it to give up and escape a life that they crazily call 'happy'. The very word 'escape' admits how weak their position is as evidenced by the example of the well-known suicide of Cato.

Augustine next tackles the Peripatetics and the Old Academics who, he alleges, call this life happy because if things become too miserable, they say they have the freedom to escape from it. But this position, he maintains, is absurd, because the happiness of this life then depends upon one's freedom to leave it, thus making its happiness congruent with the brevity of its wretchedness. Great is the power in these evils that force even the wise man to rob himself of his existence as a man, since they say that the first and greatest command of nature is that a man should be reconciled to himself and, as a consequence, naturally shun death; and that, as a living creature, he should be such a friend to himself and so wish to live in this union of body and soul that he would make continued existence his aim. The power in these evils must be great, since it overcomes this natural feeling that causes us to use all our strength in our efforts to avoid death. It so thoroughly defeats nature that what was avoided is now desired and sought, and if it is not achieved in some other way, is inflicted on a person by himself. Great, indeed, Augustine exclaims, is the power in these evils that make the virtue of fortitude a homicide, if it is proper to call fortitude that which is so thoroughly overcome by these evils that it cannot safeguard one from killing himself but rather drives him to it. The wise man ought to endure death patiently, death, that is, which he does not inflict upon himself. If, indeed, a man is compelled by these evils to kill himself, surely these philosophers must admit that they are not only evils but, in fact, intolerable evils.

Such a life, weighed down by such evils, should by no means be called happy. The men who call it that, when they are defeated by the increasing burden of ills and then surrender to adversity by killing themselves, should instead surrender to truth and believe that enjoyment of the supreme good is not a goal to be attained in our mortal state. For our virtues bear witness to life's miseries most eloquently when they support us in the midst of life's dangers, toils, and griefs. If our virtues are true, they do not claim to possess the power to ensure that we will not experience any miseries. True virtues are not guilty of such mendacity. But true virtues do say that our human life, although it must be miserable owing to the many evils of this age, *is* happy in expectation of the future age, but only if it is grounded on salvation. Hence the apostle Paul, speaking about men who live in accord with true piety, says, "Now we are saved by hope. But hope that is seen is not hope. For how would one hope for that which he sees? But if we hope for that which we do not see, we wait for it with patience" (Rom. 8:24–25). Augustine concludes this, his final discussion of suicide, by asserting that just as Christians have been saved by hope, so also have they been made happy by hope. Just as Christians do not already possess a present salvation but look forward to a future salvation, so also it is with their happiness. All must be *per patientiam*. Christians are afflicted with various evils but they must patiently endure them until they come to heaven where they will be made ineffably happy.

Such is the salvation that will be in the future age, which will itself also be the ultimate happiness. Philosophers, since they do not see this happiness and, accordingly, are not willing to believe in it, try in this life to counterfeit for themselves the falsest kind of happiness by a virtue that is as much more arrogant as it is more fraudulent.

Augustine's most frequently cited discussion of suicide, which is also his earliest thorough treatment of it,[46] is his digression on suicide in Book One of the *City of God*, which he could not have intended to be a systematic theological, much less philosophical, analysis of the subject. Augustine entered the discussion convinced that suicide was a reprehensible sin and a crime. He was speaking to his fellow Christians, who would not have been particularly surprised by anything that he was saying, because they held the same basic values as Augustine and his older contemporaries, who had already penned condemnations of suicide. The very fact that, in his discussion of suicide, his general condemnations of the act are almost incidental to his main line of argument, is in itself compelling evidence for the stability of a tradition of condemnation of suicide. Indeed, if it were not for the exception that a few church fathers had made of some virgins 'martyred' by their own hand to preserve their chastity in the face of persecution, the subject of suicide would almost certainly never have come up in Book One of the *City of God* and the only discussion of it in this work would then be that which appears in Book Nineteen.

It is also unlikely that Augustine would have used the terms that he did to describe the immorality of suicide if his Christian audience would have found his emphatic and unequivocal condemnation of the act at all remarkable. His first reference to suicide is to the suicide of Christian women in Rome who chose death in lieu of possible sexual violation by the conquering Goths. He calls their act a *facinus*. He also refers to suicide as a *scelus*, in fact a *detestabile facinus et damnabile scelus*, a *peccatum*, indeed a *peccatum gravissimum*, a *crimen homicidii*. Augustine refers to one who commits it as a *homicida*. The deed is *non licet*; it is *nefas*, a deviation *ab auctoritate legis divinae*, for which *nulla causa iusta* is possible. It is, in Augustine's mind, so obvious that the immorality of suicide would be self-evident to Christians that he can say that *veritas manifesta* itself proclaims that suicide is a *detestabile facinus et damnabile scelus*. So certain is he of the agreement of the Christian reader that he declares to believe otherwise than that Samson's self-destruction was in response to a direct command from God's Spirit *fas non est*.

Augustine based his condemnation of suicide most fundamentally on the same presuppositions and values that caused the earlier church fathers to condemn the act. His position is more developed than theirs. Earlier sources simply *assumed* that suicide was a sin. Recall the words with which they describe it: it is opposed to the will of God (Justin); suicides are punished more severely than others (*Clementine Homilies*); it is not permitted (Clement); we all justly regard suicides with horror because God punishes the murderer of self more than he does the murderer of another (John Chrysostom); nothing can be more wicked than suicide (Lactantius); Christ will not receive the soul of a suicide (Jerome); Scripture forbids a Christian to lay hands on himself (Ambrose). These are all simply assertions with which the earlier authors who made them were so confident that their Christian readers would be in agreement that they felt no need to defend them. The last example, that Scripture forbids a Christian to lay hands on himself, was from Augustine's mentor Ambrose, who gives the suicide of a woman to preserve her chastity as the only exception to this ostensibly 'scriptural' prohibition. It is, of course, the question of this exception that precipitates Augustine's digression on suicide in Book One of the *City of God*.

The only difference between Augustine and his predecessors who deal with suicide is that in Book One of the *City of God* he attempts to answer possible objections to the traditional condemnation of suicide. His defense of the traditional position is fourfold:

1. Scripture neither commands it nor expressly permits it, either as a means of attaining immortality or as a way to avoid or escape any evil.
2. It must be understood to be forbidden by the sixth commandment.
3. If no one on his own authority has a right to kill even a guilty man, then one who kills himself is a homicide.
4. The act of suicide allows no opportunity for repentance.

There are three major themes in Augustine's anti-Donatist writings that do not appear in his digression on suicide in Book One of the *City of God*:

1. Provoking martyrdom is a form of suicide and hence a sin. Persecution

by pagans was a moot point by this time. Consequently, Augustine's treatment of voluntary martyrdom, since it involves those outside the orthodox community who were responding to efforts by the Catholic Church to compel them to enter the orthodox fold, differs somewhat in force and emphasis from earlier condemnations of voluntary martyrdom by members of the orthodox community.

2. Heroic suicides by these same folk, when they were unable to provoke others to martyr them, has no known parallel in earlier Christian experience. While the act and circumstances are novel, Augustine's condemnation of such suicides is entirely consistent with the position that he had already articulated in Book One of the *City of God*.
3. In one of his last anti-Donatist treatises (*Letter* 204, written in 420), Augustine argues that the Donatists' suicides violate the foundational, Christian principle of patient endurance. Already, in his treatise on patience, written five years earlier, he had condemned suicide precisely on these grounds. In his final discussion of suicide in Book Nineteen of the *City of God*, written about a decade later, he uses patient endurance as the central principle for a condemnation of suicide as encouraged by various pagan philosophical schools.

Did Augustine formulate the Christian position on suicide? No; but he is the first Christian to discuss it thoroughly. Did he contribute anything new to the Christian position? Here again the answer is "no", except for his unequivocal condemnation of suicide to preserve chastity. It is unlikely that any of the earlier sources that I have cited would have disagreed with any of his arguments, except regarding suicide to preserve chastity, especially virginity. These authors were intelligent men who based their conclusions on their own understanding of Scripture, which they regarded as providing fundamental truth and the only basis for establishing their values and ethics. That none of the extant sources prior to Augustine had recourse to the sixth commandment as forbidding suicide is not remarkable, since earlier sources did not so much argue against suicide as assume its essential sinfulness. Granted, it is interesting historically, as a matter of record, to identify the earliest use of the sixth commandment for this purpose. But it would be significant in a discussion of the development of the Christian position on suicide only 1) if sources prior to Augustine had directly or indirectly approved of suicide generally, thus making him the earliest surviving Christian source to condemn the act; or 2) if earlier Christian authors who did condemn suicide had attempted to justify their position from Scripture (which they did not) but had failed to base any part of their argument on the sixth commandment. Neither of these, of course, is true.

There is no evidence that, at any time during the centuries under consideration, suicide was a debated issue in the Christian community. The church fathers were anything but shy about condemning moral laxity and the diverse sins of their Christian audiences. Indeed, many of them warmed to that task with an enthusiasm of moral vigor, and they dealt not only with matters specifically prohibited by Scripture but with *adiaphora* (grey areas) as well. Nevertheless, one never finds an exhortation to refrain from suicide even in the writings of those authors who condemn the act in no uncertain terms, owing simply to the fact that it was not an option for Christians.[47] The absence of a debate over suicide in the literature of early Christianity is not an indication that the Christian community was indifferent to it as an ethical issue, but rather that its condemnation as a sinful act was generally assumed throughout the period under consideration.

AFTER AUGUSTINE

Augustine is typically credited with formulating the subsequent "Christian position" on suicide and, by doing so, effecting a considerable change in the attitude of the Christian community, which is said to have turned Christianity from a gentle condoning or even a strong support of suicide to a vitriolic condemnation of the act. The evidence shows that this claim simply is not true. What is especially intriguing, indeed ironic, is that in the only matter relevant to suicide in which Augustine disagreed with any earlier or contemporary church fathers (women committing suicide to preserve their chastity), he did not carry the day. M.P. Battin, who maintains that "there is little reason to think that Augustine's position is authentically Christian" ([8], p. 71), compounds her error by saying that support for the earlier position regarding suicide to preserve chastity "ends, however, with St. Augustine's repudiation of suicide as an alternative to sexual defilement..." (*ibid*, p. 164, n. 16). Here Battin contradicts herself, since earlier she had quite correctly asserted that there is "no doubt that the principle [*sc.* of double effect] has been applied in some very strained ways. A particularly glaring example is the *traditional justification* of suicide by women to protect their virginity..." (my emphasis, *ibid*, p. 66). The "traditional justification" of such suicides that she has in mind is not the one to which she later claims that Augustine put an end. Rather, justification of such suicides is a tradition that is still quite vital, as she demonstrates by citing volume two of *Moral and Pastoral Theology*, written in 1936 by the Jesuit Henry Davis ([13], vol. 2, p. 116).

Numerous examples could be given here of theological justifications of such suicides by Roman Catholic casuists and moral theologians from the Middle Ages to the present. One very influential work must suffice as an illustration: Hermann Buzembaum's *Medulla Theologiae Moralis*. First printed in 1648, it has enjoyed numerous reprintings. According to Buzembaum, it is "not deemed suicide ... if a virgin, to preserve her chastity, were 'to embrace even certain danger of death', so great a good was her 'integrity'" ([32], p. 124). Suicide by virgins to preserve their chastity has remained a grey and disputed area and, by the casuistic sleight-of-hand known as double effect, is then not regarded as suicide at all. And it has continued to be morally defended, in spite of such greats of Catholic tradition as Augustine and Aquinas. Thomas Aquinas (1225–1274) stands even higher than Augustine as an authoritative figure in Roman Catholicism. In his *Summa Theologiae*, while discussing martyrdom (2a2ae.124.1), he condemns suicide by virgins to preserve their chastity. His main discussion of the act, however, occurs earlier, in his section on homicide (2a2ae.64.6), which contains his classic treatment of suicide. Here he follows Augustine very closely, allowing no exceptions to the condemnation of suicide (except the hypothetical case of a direct, personal command of God). Aquinas' argument, which is significantly shorter and much more concise than Augustine's various treatments of the subject, has been closely scrutinized by numerous philosophers and theologians. Aquinas' position is based upon three "reasons" why it is "altogether unlawful to kill oneself": 1) It is unnatural, that is, contrary to the inclination of nature, for everything seeks to keep itself in being, and it is contrary to charity because every man should love himself. 2) It is an offense against, and an injury to, one's community, because one belongs to his own community. 3) It is a usurpation of God's power to give and take away life.

Aquinas' argument against suicide begins with a quotation of Augustine's application of the sixth commandment to suicide (*City of God* 1.20), after which he gives the three already stated reasons regarding suicide as *omnino illicitum*. The remainder of his discussion is an elaboration of these three "reasons". Suicide is contrary to justice and to "that charity which a man owes to himself" and is therefore a sin. It is "a sin against justice when considered in relation to the community and to God". He then distinguishes between lawful killing of a malefactor and suicide, asserting that "nobody is a judge in his own cause".

Aquinas' next section begins with the assertion that free will makes one a master of himself. One's free will is limited, however, to the management of one's life. Hence, since the passage from this life to "a more blessed one" is not the province of man's free will, one may not kill himself. The same applies to suicide when committed 1) to escape from the miseries of this life; 2) to punish oneself for having committed some sin; 3) to avoid (in the case of a woman) being sexually violated; and 4) to avoid a sin that one is in fear of committing. While considering 2 and 3, Aquinas stresses that suicide deprives one of an opportunity for repentance.

Aquinas then deals very briefly with the problem presented by the suicides of Samson and those virgins "held in memory by the Church" (*quarum memoria in Ecclesia celebratum*) who committed suicide to preserve their chastity. He simply quotes Augustine's assertion that they must have acted under the command of the Holy Spirit. He concludes his discussion of suicide by citing Razias as an example of the "sort of courage" which is really cowardice rather than true courage which is the ability to bear afflictions. He cites both Aristotle and Augustine as authorities for this conclusion.

What, if anything, new did Aquinas add to the traditional Christian arguments against suicide? Battin maintains that two arguments originated with Aquinas: "1) that suicide is forbidden because it attempts to usurp God's judgment over 'the passage from this life to a more blessed one'; and 2) the warning that suicide is 'very perilous' because it leaves no time for repentance" ([8], p. 52). She is clearly wrong in regard to the latter, for Augustine, as we have seen, emphasizes that very point. In regard to the former, she is technically correct. Aquinas appears to have been the first to employ this argument, articulated in this precise way, in condemning suicide. However, a fervent belief in the sovereignty of God undergirds Augustine's entire *City of God* and conditions his discussion of suicide throughout.

Aquinas originated neither of the other two "reasons" that he gives for the absolute illicitness of suicide, although his use of them as two of only three foundational principles for the condemnation of suicide is new. In reference to the first of these, we should note that in Book Nineteen of the *City of God*, Augustine agrees with those philosophers who maintain that the first and greatest command of nature is to shun death and make continued existence one's aim. His conclusion is that the compulsion to end one's life thoroughly defeats nature. Also, in *De libero arbitrio*, he uses the assertion, as an oblique argument against suicide, that existence is *eo ipso* a good and that even a miserable and unhappy existence is better than non-existence. He further maintains that no one should be ungrateful to the goodness of his Creator for his existence (n. 46, below).

Aquinas' second reason, for which he cites Aristotle as his authority, is an argument that would not have seemed particularly compelling during Augustine's time when civilization was tottering on the brink of destruction. But it was a central verity of the late medieval ethos. The highly corporate structure of life in Aquinas' time fostered a deontology predicated upon community. It is not at all remarkable that Aquinas included the principle of community in his treatment of suicide. It was thoroughly Aristotelian. So was Aquinas. And it was, in the most vivid of ways, foundational for contemporary concepts of obligations. It would truly have been remarkable if Aquinas had not had recourse to this principle as an essential element in his argument against suicide; it would have been equally remarkable if Augustine had.

Regarding their treatment of suicide one can say, add Aristotle to Augustine, and the end product is Aquinas.

There are two matters that neither Augustine nor Aquinas considered in their discussions of suicide, both of which were matters of interest to some members of late classical (but Christianized) society and the medieval community. One of these involves mitigating or extenuating circumstances affecting the suicide. The one mitigating or extenuating factor was insanity/demon possession. In late antiquity Chrysostom ([33], pp. 77–78) and John Cassian (*Conferences*, 2.5) dealt with a phenomenon known in Greek as *athumia* and in Latin as *acedia*, a condition subsumable under the catch-all rubric of melancholy, to which monks were especially susceptible. Numerous examples of suicides or of contemplated or attempted suicide owing to this condition are preserved by medieval sources. Early medieval penitentials wrestled with suicides by the insane/demon possessed. A good example is the Penitential of Theodore (ca. 668–690):

> *Of Those Who Are Vexed by the Devil*: If a man is vexed by the devil and can do nothing but run about everywhere, and [if he] slays himself, there may be some reason to pray for him if he was formerly religious. If it was on account of despair, or of some fear, or for unknown reasons, we ought to leave to God the decision of this matter, and we dare not pray for him. In the case of one who of his own will slays himself, masses may not be said for him; but we may only pray and dispense alms. If any Christian goes insane through a sudden seizure, or as a result of insanity slays himself – there are some who celebrate masses for such one ([28], p. 207).

Here we enter into a quagmire of confusion that demands a thorough analysis well beyond the scope of the present essay. In a previous study I suggested that, since

> it is so common to assume that in the early Middle Ages demonic causality was

accepted for most disease, and certainly for mental illness, such natural causes as despair, fear, sudden seizure, and "unknown reasons" may surprise us. Such explanations, however, occur with some frequency.... There is, however, considerable ambiguity in the literature. This ambiguity arises from two different sources. First, there was much imprecision in identifying the causes of mental illness. The second source of ambiguity is the failure of the modern reader to enter sufficiently into the early medieval structure of reality in which ultimate and proximate causality may be spoken of in the same breath without any distinction being made, and an intermediate (usually demonic) causality mingled in with the former two. Any one of the three may be mentioned as the cause of a particular condition, and taken by the modern reader as the author's perceived sole cause, whereas the choice of that cause was simply determined by the author's desire to emphasize one with no intention of making it appear exclusive ([2], pp. 75–76).

It must be sufficient that the reader be aware that during the period between Augustine and the Renaissance and Reformation insanity/demon possession rendered suicide not as clear an issue as we might assume based on some modern discussions of that period. This held true not only in the ecclesiastical realm but in secular law as well. Henry of Bracton, an English jurist living in the thirteenth century, for example, exempted suicide by the insane from criminal considerations ([39], p. 13).

The second matter with which neither Augustine nor Aquinas dealt that was a concern to the medieval community is the question of temporal penalties against the suicide. While ecclesiastical law and practice involved questions of sacraments for the deceased and proper manner of burial, secular law went much further. When considering the latter, we enter into a chamber of horrors, including such customs as, for example, exposing the suicide's body at crossroads with a stake driven through his heart and confiscation of all his temporal goods to the obvious detriment of his family.

Augustine may have known of pagan customs of dishonoring a suicide's corpse. It is possible that he was aware of some classical practices that have been preserved in the literature. It is less likely that he was acquainted with contemporary practices in Europe. He advocates nothing similar. Nor does he suggest anything like the denying of sacraments and burial rites to suicides, policies adopted by regional church councils throughout the Middle Ages. Glanville Williams, after citing the action of the Synod of Nimes in 1284 that "refused suicides even the right of quiet interment in holy ground", comments:

This canonical development was undoubtedly influenced in part by the writings of St. Augustine; but he had not demanded any punishment for the deed. In fact, as has been seen, Augustine was prepared to justify some historic suicides as specially commanded by God, and, consistently with this theory, he felt himself constrained to admit that even in his own day a person might rightly make away with himself by special command of God: "only let him be very sure that the divine command has been signified". Such a theory is obviously inconsistent with the anathematizing of the suicide's corpse by way of punishment, for what ecclesiastic can say, after the event, whether the divine command was signified or not? There is another reason for absolving Augustine from sole responsibility for these laws. The severe penalty of depriving burial rites smacks of the pagan practice of dishonouring the corpse, and Bayet consequently maintains, with much force, that it represents an irruption into the church of the pre-Christian popular horror of suicide. Like so much else in ecclesiastical practice and belief, it is a pagan intrusion upon the simple philosophy of the Gospels ([43], p. 258).

Even this absolving of Augustine of responsibility for ecclesiastical and secular practices that were common during the Middle Ages and much later is inadequate, because it hinges on the false premise that Augustine introduced something new into the Christian attitude toward suicide. Williams is correct in seeing some medieval and later practices as "a pagan intrusion upon the simple philosophy of the Gospels", but he demonstrates a lack of discernment of the nature of this "simple philosophy of the Gospels" by attributing attitudes to the early Christian community that are inconsistent with the values consistently expressed both in the New Testament and in patristic literature. Furthermore, he sees pagan intrusions into Augustine's thought:

Augustine's third argument was stoical in conception, though he rejected the Stoics. The truly great mind, said Augustine, will bear the ills of life. The argument reappears from time to time in the proposition that suicide is an act of cowardice. Now, the only line between cowardice and caution (or wise retreat) is that the coward does not do what he ought to do. To brand the suicide as a coward is, therefore, to beg the question whether there is a duty to go on living (*ibid.*, p. 256).

This quotation reveals a misunderstanding, common to many modern scholars, of the most fundamental principles of Christian thought that permeate the New Testament and patristic literature at least through the age of Augustine. And these fundamental principles of Christian thought encourage patience, stress hope, command perseverance, and so strongly emphasize the sovereignty of God in the life of Christians and the trust that they should have, based on the conviction that God will cause all things to work together for their ultimate good, that it is unlikely that suicide for such reasons as Augustine condemns would have been regarded by the vast majority of orthodox Christians before and during his time as anything less than a breach of trust. The one significant exception is the suicide of women, especially virgins, to preserve their chastity. And it is ironic that it is in this area that Augustine continues to be at variance with some Catholic moral theologians.[48]

NOTES

[1] Ex. 21:22–25, the one possible exception, is patient of two different interpretations: the "serious injury" applies 1) to the mother or 2) to the baby. Regardless, the passage does not involve intentional, induced abortion.
[2] We now recognize, however, that they often confused abortion and contraception owing to their rudimentary knowledge of obstetrics and embryology.
[3] For a thorough discussion, see [12].
[4] E.g., Mt. 8:22; Lk. 15:32; Rom. 5:12–19; 6:23; Eph. 2:1, 5:14; Col. 2:13; Jas. 1:15.
[5] Quotations from the New Testament are taken from the Revised Standard Version.
[6] The New Testament teaches that Christ has destroyed spiritual death for the believer (2 Tim. 1:10; 1 Cor. 15:25–26; Heb. 2:14). Spiritual death is the opposite of salvation or eternal life, the latter contrasted ultimately with the 'second death' which is the eternal state of the unregenerate (Rev. 2:11; 20:6 and 14; 21:8; *cf.*, Mt. 25:41 and 46; Jn. 3:36; 5:28–29).
[7] E.g., Mt. 13:43; Lk. 10:20; Jn. 14:2–3; 17:24; Rom. 8:17; 1 Cor. 15:12–19; 2 Cor. 4:17; Eph. 1:18; 1 Thess. 4:13–18; 1 Titus 2:13; 1 Pet. 1:3–4; Jude 21; Rev. 21:1–4.
[8] E.g., Jas. 4:13–16; 1 Pet. 1:24.
[9] = Mt. 16:25–26; *cf.*, Mt. 6:19–21.
[10] Phil. 3:20; Heb. 13:14; 1 Pet. 2:11.
[11] E.g., Jn. 10:27–30; Rom. 6:1–10; Eph 2:6; Col. 2:12; 3:1.
[12] *Cf.*, Mt. 13:43; Jn. 14:3; Rom. 8:17; 2Cor. 4:17.
[13] E.g., Mt. 27:52; Jn. 11:11–14; Acts 7:60; 13:36; 1 Cor. 15:6, 18, 20, and 51; 1 Thess. 4:13–15.
[14] *Cf.*, Lk. 23:43; Acts 7:59–60.
[15] *Cf.*, Jn. 14:21 and 23; 15:10.
[16] = Mk. 12:28–31 = Lk. 10:25–37.
[17] See, e.g., Jesus' specific injunction, Mt. 28:19–20; Acts 1:8; *cf.*, Acts 10:42.
[18] Jn. 15:20; *cf.*, Mk. 10:28–30; Acts 9:16; 14:22; 1 Thess. 3:2–4; 2 Tim. 1:8; 2:3; 4:5.
[19] *Homologesēi en emoi* is an Aramaism for "confesses me".
[20] E.g., Jn. 17:4–5; Phil. 2:8–9; Heb. 2:9.
[21] *Cf.*, 1 Pet. 1:6–7.
[22] Mt. 11:29; Phil. 2:5–8; 1 Pet. 2:21.
[23] *Cf.*, 2 Cor. 11:32–33.
[24] Somewhat in the same vein 1 Timothy urges its addressee to pray for "kings and all who are in the high positions, that we may lead a quiet and peaceable life" (1 Tim. 2:2).
[25] *Cf.*, Mt. 5:48; Eph. 4:22–5:14; Col. 3:5–10.
[26] E.g., Acts 13:38; Eph. 1:7; Col. 1; 14; Heb. 9:14; Jas. 4:8; 1 Jn. 1:9; 3:3.
[27] This was a point of considerable dispute during the early centuries of Christianity

as views varied enormously on the nature of forgiveness of post-baptismal sins and the state of those who had lapsed during times of persecution and then wished to be re-admitted to the Christian community.

[28] *Cf.*, 1 Tim. 4:3.

[29] = Mk. 13:13; *cf.*, Mt. 24:13.

[30] E.g. Rom. 11:22; Heb. 3:6 and 14; Rev. 2:7, 11, 17, and 26; 3:5, 12, and 21; 21:7.

[31] The writer of Hebrews is here quoting Prov. 3:11–12. The words translated 'discipline', both in the Hebrew of Proverbs as well as in the Greek of this epistle, mean discipline in the sense of training, instruction, and upbringing.

[32] *Cf.*, 2 Tim. 1:11–12; 1 Pet. 4:19; 5:6–7.

[33] For an insightful discussion, see [11].

[34] E.g., Pseudo-Justin, *Epistle* 17; Hippolytus, *Apostolic Tradition*, canon 20; Polycarp, *Epistle* 6; Pseudo-Clement, *De Virginitate*; Tertullian *Ad Uxorem* 2.4; Justin, *Apology* 1.67; Jerome, *Epistle* 52.15 and 16.

[35] There is an enormous literature on the subject of martyrdom and persecution in early Christianity. The most authoritative and reliable treatment is [22].

[36] For a discusion of this case, see [7], pp. 168–171.

[37] What may have been more common were Christians provoking persecutions and potential martyrdom by symbolic acts such as destroying pagan idols. Canon 60 of the Synod of Elvira, held around 305, condemned such acts on the ground that the gospel does not suggest deeds of this kind and the apostles did not act in this fashion. The synod decreed that people who died as a result of doing so should not be regarded as martyrs ([25], p. 163).

[38] See, as a beginning point, [14], especially pp. 26–36.

[39] E.g., Clement of Alexandria, *Stomateis* 3.7; Lactantius, *The Divine Institutes* 5:22; Ambrose, *Epistle* 63:91.

[40] See [29].

[41] Somewhat similarly, Tertullian gives many examples of pagan men and women who did not shrink from horrible sufferings. Some even sought suffering for the sake of glory. Others, "led by the impulses of their own mind, put an end to their lives". Under this rubric he includes the Roman matron Lucretia; the legendary Carthaginian queen, Dido; the wife of Hasdrubal, the Carthaginian general who surrendered to Scipio; Cleopatra; and the philosophers, Heraclitus, Empedocles, and Peregrinus who, after a brief stint as a Christian, became a Cynic and ended his life by self-immolation during the Olympic Games in 165 (*To the Martyrs* 4–9 [19], vol. 40, pp. 24–28).

Tertullian's assessment of these suicides and of those who sacrificed their lives or bravely endured horrible sufferings is that, "if earthy glory accruing from strength of body and soul is valued so highly" that pagans would undergo such things "for the reward of human praise", then the sufferings that Christian martyrs endure "are but trifling in comparison with the heavenly glory and divine reward" (*ibid.*, 9, pp. 27–28).

In his *Apology* he compared Lycurgus, who had "hoped to starve himself to death because the Spartans had amended his laws", with the Christian, who, "even when condemned, gives thank" (46.14 [19], vol. 10, p. 113). Later he mentions Empedocles' and Dido's self-immolation. Here he becomes sarcastic: "Oh, what strength of mind!" he exclaims about the former, and about the latter, "Oh, what a glorious mark of chastity!". He says that he will "pass over those who by their own sword or by some other gentler manner of death made sure of their fame after death". On these suicides and on certain examples of bravery and fortitude, he comments that "such recklessness and depravity, for the sake of glory and renown, raise aloft among the banner of courage". But, in the opinion of the pagans, the man who "suffers for God, is a madman" (*ibid.*, 50.4–11, pp. 123–125).

Tertullian does not condemn these pagan suicides, at least not directly. His sole purpose, as Timothy Barnes expresses it in his study of Tertullian, is contrast: "If glass is so precious, how valuable must be a genuine pearl! Why should Christians hesitate to die for the truth, when others die for false ideals such as their own glory?" ([7], pp. 218–219). "These are all pagans with false ideals. How much more should the Christian endure for the sake of truth and for celestial glory" (*ibid.*, p. 227). Tertullian's sarcasm shows his readers what he thinks of these suicides. If pagans are willing to kill themselves for such unworthy reasons, how much more understandable it is that Christians willingly die for their faith.

[42] Chrysostom, in his *Homilia Encomastica* on the most famous of these virgin suicides, Pelagia (who died about 313), enthusiastically approved her act ([34], vol. 49, cols. 579–584).

[43] All subsequent quotations from the *City of God* in the present paper are my own translations.

[44] The only examples of Christians committing suicide to avoid torture are those during the Great Persecution mentioned by Eusebius (8.12.2). Eusebius expresses neither approval nor disapproval. This type of suicide must have been very uncommon since it attracted so little attention by the sources. It is likely that it was so rare that Augustine was unaware of specific examples. He was so thorough in his various discussions of suicide that it is most improbable that he would have missed such incidents if they were well known. And if he had been aware of them, it would be totally uncharacteristic of him to have shied away from wrestling with the possible problems posed by such cases. He will later address the subject when considering the Donatists, a schismatic, heretical sect.

[45] For a thorough, scholarly treatment of the Donatist movement, see [21].

[46] The earliest mention of suicide in Augustine's writings is in his *De libero arbitrio* or *The Free Choice of the Will*, published in three books between 388 and 395. Augustine was converted to Christianity in 386. The *De libero arbitrio* was the eighth and last in a series of dialogues of a predominantly philosophical nature. The subject matter of the first seven includes the problem of certitude, human happiness, providence and the problem of evil, God and the soul, immortality, and the function of language and the role of the teacher in learning. This was the philosophical period of Augustine's literary output, a time when he and his small circle of interlocutors sought to clarify a wide variety of issues by the use of reason illumined by Christian revelation, striving to preserve their intellectual integrity while remaining consistent with Christian truth, in the broadest sense, as they conceived it.

Augustine composed the *De libero arbitrio* as a refutation of the deterministic dualism of Manicheism, the sect from which he had recently departed. The work is in great part a theodicy that attempts to establish the existence and goodness of God, the freedom of the created will, and the latter as the sole and adequate cause of moral evil. There is nothing conspicuously Christian about most of the work. Indeed as an old man, Augustine explained in his *Retractations* that he had mentioned God's grace so little in this work because his major concern in writing it was to argue for human responsibility against a Manichean determinism that denied it. It is only as the work is reaching its conclusion that he connects the theme of freedom of the will with such doctrines as original sin and salvation through Christ's atoning sacrifice.

Well into the final book of the *De libero arbitrio* Augustine asserts that existence is *eo ipso* a good and that even a miserable and unhappy existence is better than non-existence. He deals with a variety of types of people who are wretchedly miserable but choose to go on living even though they claim to prefer extinction. The unhappiness of such people is just, since they are unwilling to praise God and be thankful for the good of existence itself. He then exclaims:

[H]ow absurd and inconsistent it is for anyone to say that he would rather not exist than be unhappy! A man who says he would rather have this than that is making a choice of something, whereas non-existence is not something, but nothing. It is logically impossible, therefore, for you to make a choice when the object of your choice does not exist.

He maintains that one should not wish for non-existence since that which does not exist cannot be better than that which does exist. He ends this line of reasoning with the statement that "when a man has attained what he rightly chose as something to be desired, he necessarily becomes a better man. But he cannot become better if he is not going to exist. No one, therefore, can logically choose not to exist." At this point he turns to the question of suicide:

Nor should we be unsettled by the judgment of those men who have been driven by unhappiness to take their own lives. *Either they have sought to find refuge where they imagined they would be better off, and this view, however they may have come to it, is not opposed to our line of reasoning* [my emphasis]; or if they thought they would no longer exist at all, we will be far less unsettled by the illogical choice of men who make "nothing" the object of their choice. If anyone chooses non-existence, he is obviously choosing "nothing", even though he is unwilling to give this for an answer. Allow me, however, to express my own view, if I can, about this whole question. In my opinion, no one who takes his life or in any way desires death, really feels certain that he will not exist after death, even though he holds it somewhat as an opinion.

Now, after distinguishing between opinion, reason, belief, and feeling, he concludes that,

when anyone is driven by unbearable hardships to desire death wholeheartedly, in the belief that he will not exist after death, he decides upon death and grasps for it. Opinion leads him to entertain the false notion of a complete extinction, whereas feeling suggests a natural desire to be at rest. But the state of rest is not the same as nothing; on the contrary, a thing at rest exists more perfectly than that which is not at rest. In fact, restlessness makes us vacillate in our affections so that one of them destroys the other, while rest possesses a constancy which is uppermost in our mind when we say of anything that it exists. Accordingly, every desire on the part of a man's will for death is directed, not towards extinction after death, but towards rest. Though he has the mistaken belief that he will not exist, he still has a natural desire to be at rest, that is, to enjoy a more perfect existence. Hence, just as no one can possibly find delight in not existing, so *it should never happen that anyone should be ungrateful towards the goodness of his Creator for his existence* [my emphasis] (*De libero arbitrio* 3.7.20–3.8.23 [19], vol. 59, pp. 182–187).

I am not concerned with attempting to assess the validity of Augustine's argument here, but simply with pointing out three matters of significance:

1. Augustine's purpose here was not to condemn suicide as a moral wrong, but rather to argue that existence is a greater good than non-existence as part of a theodicy that focuses on the freedom of the will.
2. Hence, the first italicized section above cannot reasonably be taken as indifference toward or approval of suicide when motivated by the hope of a better existence. We can assume that neither the orthodox community nor heretical sects took it that way, for if they had, Augustine would have corrected that misunderstanding as he did several other matters in the section of his *Retractations* devoted to the *De libero arbitrio*. Furthermore, it can be argued that Augustine implies a condemnation of suicide *if* one combines the assertion in the last italicized clause above with Augustine's argument in the first book of the *De libero arbitrio* that although human laws allow one to kill in defense of property and of his own life, the unseen law of Divine Providence does not: "How, in the light of this law, are they without sin who defile themselves by human slaughter for the sake of things which ought to be despised?" (1.5.13, *ibid.*, p. 83). If one sins and defiles himself by killing another for the sake of temporal things that ought to be despised, including life itself if the only way it can be preserved is through killing another, so also would one sin and defile himself by suicide when motivated by an unfulfilled desire for the temporal good of happiness.
3. There is an important similarity between Augustine's including a brief discussion of suicide in his *De libero arbitrio* as a subordinate argument and his digression on suicide in Book One of the *City of God*. That similarity is that neither one was intended to be a systematic and comprehensive treatment of the moral issues involved in a consideration of suicide. Both discussions of suicide must be appreciated within the context of Augustine's immediate purposes, in his formulation of a position to which a consideration of suicide is only incidental.

[47] The only hint that there might be even a minor problem is in the *City of God* 1.25, where Augustine says that he included the discussion of the question of suicide committed in order to avoid sinning, "for the sake of these men or women who think that they must do mortal violence to themselves in order to avoid, not another's sin, but a sin of their own". When he returns to the subject in 1.27, he concedes that, "if any just cause were possible for suicide, I am sure that there could not be one more just than that". But we should not take this as anything more than an acknowledgement of the realities of the spiritual warfare that caused even Paul, in his yearning to be entirely free from sin, finally to cry out, "Wretched man that I am! Who will deliver me from this body of death?" (Rom. 7:24). The scenario of the discouraged Christian, who, vitally concerned about this propensity to sin, asks his spiritual

counsellor why he can not just end the battle by going to heaven immediately, is more likely than one in which a person who wants to kill himself argues that this might be a justifiable cause.

This is not to say that there are no cases of newly converted Christians committing suicide with the intention of going directly to heaven. But the earliest examples that I have been able to locate come from the British Isles and survive in documents written centuries after Augustine's time. See [33], pp. 75–76).

[48] I wish to express my appreciation to Dr. Gary B. Ferngren for his astute criticism of the penultimate version of this paper, and to Valerie Worthen for her patience in turning numerous pages of nearly illegible scratchings into clean copy.

BIBLIOGRAPHY

[1] Alvarez, A.: 1980, 'The Historical Background', in [9], pp. 7–32.
[2] Amundsen, D.W.: 1986, 'The Medieval Catholic Tradition', in [31], pp. 65–107.
[3] *Ante-Nicene Fathers*: various dates, various translators, rpt. 1951, Eerdmans, Grand Rapids, Michigan.
[4] *Apostolic Fathers*: 1912–13, trans. by K. Lake, Harvard University Press, Cambridge.
[5] Augustine: 1972, *City of God*, trans. by H. Bettenson, Penguin, London.
[6] Barnard, L.W.: 1967, *Justin Martyr: His Life and Thought*, Cambridge University Press, Cambridge.
[7] Barnes, T.D.: 1971, *Tertullian: A Historical and Literary Study*, Clarendon Press, Oxford.
[8] Battin, M.P.: 1982, *Ethical Issues in Suicide*, Prentice Hall, Englewood Cliffs, New Jersey.
[9] Battin, M.P., and May, D.J. (eds.): 1980, *Suicide: The Philosophical Issues*, St. Martin's Press, New York.
[10] Brown, P.: 1981, *The Cult of the Saints: Its Rise and Function in Latin Christianity*, University of Chicago Press, Chicago.
[11] Daube, D.: 1962,'Death as a Release in the Bible', *Novum Testamentum* 5, 82–104.
[12] Daube, D.: 1972, 'The Linguistic of Suicide', *Philosophy and Public Affairs* 1, 387–437.
[13] Davis, H.: 1936, *Moral and Pastoral Theology*, Sheed and Ward, London.
[14] Dodds, E.R.: 1965, *Pagan and Christian in an Age of Anxiety*, rpt. 1970, Norton, New York.
[15] Douglas, J.D. (ed.): 1974, *The New International Dictionary of the Christian Church*, 2nd edition, Zondervan, Grand Rapids, Michigan.
[16] Dublin, L.I.: 1963, *Suicide: A Sociological and Statistical Study*, The Ronald Press Co., New York.
[17] Dudden, F.H.: 1935, *The Life and Times of St. Ambrose*, Clarendon Press, Oxford.
[18] Eusebius: 1926–32, *Ecclesiastical History*, trans. by K. Lake and J. Oulton, Harvard University Press, Cambridge.

[19] *Fathers of the Church*: 1948–, various editors and translators, The Catholic University of America, Washington, D.C.
[20] Fisher, G.P.: 1911, *The Beginning of Christianity*, Scribner's New York.
[21] Frend, W.H.C.: 1952, *The Donatist Church*, Clarendon Press, Oxford.
[22] Frend, W.H.C.: 1965, *Martyrdom and Persecution in the Early Church*, Basil Blackwell, Oxford.
[23] Getty, M.M.: 1931, *The Life of the North Africans as Revealed in the Sermons of Saint Augustine*, The Catholic University of America, Washington, D.C.
[24] Harnack, A.: 1905, *The Expansion of Christianity in the First Three Centuries*, trans. by J. Moffat, rpt. 1972, Books for Libraries Press, New York.
[25] Hefele, C.J.: 1871, *A History of the Christian Councils*, T. and T. Clark, Edinburgh.
[26] Labriolle, P. de: 1924, *History and Literature of Christianity from Tertullian to Boethius*, rpt. 1968, Barnes and Noble, New York.
[27] Lucian: 1968, *Selected Satires of Lucian*, trans. by L. Casson, Norton, New York.
[28] McNeill, J.T., and Gamer, H.M.: 1938, *Medieval Handbooks of Penance*, rpt. 1965, Octagon Books, New York.
[29] Miles, R.M.: 1979, *Augustine on the Body*, Scholars Press, Missoula, Montana.
[30] Mounteer, C.A.: 1981, 'Guilt, Martyrdom and Monasticism', *The Journal of Psychohistory* 9, 145–171.
[31] Numbers, R.M. and Amundsen, D.W. (eds.): 1986, *Caring and Curing: Health and Medicine in the Western Religious Traditions*, Macmillan, New York.
[32] O'Connell, M.R.: 1986, 'The Roman Catholic Tradition Since 1545', in [31], pp. 108–145.
[33] O'Dea, J.J.: 1882, *Suicide: Studies on its Philosophy, Causes, and Prevention*, Putnam, New York.
[34] *Patrologia cursus completus, series Graeco-Latina*: 1857–1866, J.P. Migne, Paris.
[35] *Patrologia cursus completus, series Latina*: 1844–1864, J.P. Migne, Paris.
[36] Perlin, S.: 1975, *A Handbook for the Study of Suicide*, Oxford, New York.
[37] Quasten, J.: 1950, *Patrology*, Vol. 1: *The Beginnings of Patristic Literature*, rpt. 1983, Christian Classics, Westminster, Maryland.
[38] Ramsey, B.: 1985, *Beginning to Read the Fathers*, Paulist Press, New York.
[39] Rosen, G.: 1975, 'History', in [36], pp. 3–29.
[40] *Select Library of Nicene and Post-Nicene Fathers of the Christian Church*, First Series: various dates, various translators, rpt. 1976–79, Eerdmans, Grand Rapids, Michigan.
[41] *Select Library of Nicene and Post-Nicene Fathers of the Christian Church*, Second Series: various dates, various translators, rpt. 1976–79, Eerdmans, Grand Rapids, Michigan.
[42] Williams, C.P.: 1974, 'Lactantius' in [15], p. 575.
[43] Williams, G.: 1957, *The Sanctity of Life and the Criminal Law*, rpt. 1970, Alfred A. Knopf, New York.

Western Washington University,
Bellingham, Washington, U.S.A.

GARY B. FERNGREN

THE ETHICS OF SUICIDE IN THE RENAISSANCE AND REFORMATION

Evolving attitudes towards suicide in early modern Europe can best be understood against the backdrop of the church's traditional opposition to any form of self-destruction. The formulation of the orthodox Christian attitude began with Augustine, who in his *City of God* (1. 17–27) provided the first extensive discussion of the morality of suicide from a Christian point of view. He viewed suicide as a violation of the sixth commandment[1] and a sin that precluded repentence. It was not justified in any circumstances, he believed, since it was a greater sin than any that it could seek to avoid. Augustine's position became essentially the mediaeval Catholic position, as later amplified by Thomas Aquinas, who suggested three additional arguments: that suicide was contrary to natural law and self-love; that it deprived society of the contribution and activity of an individual; and that it usurped the function of God (*Summa Theologica* 2–2, 64, 5). At the end of the Middle Ages the Catholic condemnation of suicide was anchored in the belief that suicide contravened the unchanging moral law governing man's relationship to God, who alone gave life and took it. Man's control over himself[2] was limited (it consisted of *usus* rather than *dominium*) and this limitation did not sanction self-murder. Civil law reflected not only the Christian belief that suicide was a sin beyond forgiveness but the repugnance attached to an act that was widely thought to be unnatural. Primitive customs that probably dated from pre-Christian times involving public desecration of the corpse (e.g., the practice in England of burying suicides at crossroads after a stake had been driven through the corpse) were practiced till the nineteenth century.[3] Suicide became a crime as well as a sin and was often punished additionally by confiscation of property and refusal of consecrated ground for burial of the corpse. It was during the sixteenth and seventeenth centuries that the traditional Christian view began to be challenged, at first obliquely but gradually with increasing boldness. In the beginning the attack took place within an assumed Christian context; eventually Christian presuppositions were discarded altogether. By the end of the seventeenth century the 'modern' view of suicide had been formulated: that suicide was not an offense against God but merely a matter of personal choice unencumbered by theological or dogmatic considerations and devoid of blame or disgrace.

Baruch A. Brody (ed.), Suicide and Euthanasia, pp. 155–181.
© 1989 *Kluwer Academic Publishers.*

THE SIXTEENTH-CENTURY RENAISSANCE

The term Renaissance is widely used to denote the rebirth of interest in classical (i.e., Greek and Roman) culture that began in Italy in the fourteenth century. Its attitudes were in large part shaped by a new course of studies that was introduced into Italian universities, the *studia humanitatis*. The curriculum involved the reading of classical authors, whose values gradually came to influence the middle and upper classes of Italy. The new education later spread north as students came from all over Europe to study in Italy. Those who studied the classics tried to adapt what was best in Greek and Roman thought to their own time. Hence they sought guidance in the arts, in law and politics, and even in moral behavior. While they sometimes found a tension between Christian and classical ideas, many of the early humanists created a synthesis that incorporated both.[4] It has been maintained that the humanists, with their glorification of man and his accomplishments and their optimism, individualism, and even secular outlook, were responsible for a change of attitudes to suicide. This is the view of Sprott, Fedden, Gruman,[5] and others. "As the humanists freed men to live", writes S.E. Sprott, "they brought them the right to die" ([30], p. 1). This assessment of the influence of the Renaissance on attitudes to suicide can be accepted only with qualification. It is difficult to find much evidence of a significant change for nearly two centuries. A long tradition maintained by Stoics and others in the ancient world held that suicide provided an honorable and even praiseworthy escape from a life that had become unbearable. As widely known as this tradition was in the Renaissance, it seems to have gained few champions as late as the sixteenth century and virtually none before it. How does one account for this? Fedden attributes it largely to the exuberance of the Renaissance: a stress on human potential and *joie de vivre* as well as a desire for fame that characterized men of the time ([14], pp. 165–166). On the other hand, there was also a fatalistic and pessimistic strain in Renaissance humanism, and Aristotelians like Pietro Pomponazzi raised doubts about personal immortality. The spread of Platonism may have done something to retard acceptance of the classical justification of suicide. One of the most significant results of the revival of Greek learning in the latter half of the fifteenth century was the new interest in Plato, whose influence on Renaissance thought became enormous. Plato was generally opposed to suicide (see *Phaedo* 62), with certain exceptions, and his views may have given support to traditional Christian belief. Doubtless, however, it was the persistence of the church's condemnation of suicide that served as the strongest bulwark against the acceptance of contrary Greek and Roman views.

While the revived classical ideals, incompatible as some of them were with Christianity, might lead humanists to challenge traditional dogmas, they rarely did so. Many, like Petrarch and Erasmus, enjoyed the patronage of the church, and they often employed their scholarship in its service. Perhaps, too, because their interests lay outside theology and philosophy, they tended not to question dogma, not, at any rate, before the advent of Protestantism. The teaching of the church on matters like suicide, carefully defined for a thousand years, buttressed by the subtle discussions of generations of philosophers and theologians, maintained by the magisterium of popes and councils, and defined in canon law, was not easily challenged by the humanists, and for a long time none did. Petrarch (1304–1374), earliest of the humanists, after concluding the arguments in its favor, rejected it ([14], p. 161). So did Francisco Piccolomini, who, writing on heroic virtue over two centuries later, held that suffering was better than ending one's own life and not incompatible with virtue ([30], pp. 6–7). Abundant use was made of classical authors like Plato, Aristotle, and Pythagoras, who rejected suicide, in order to refute Stoic and Epicurean arguments in its favor. In spite of the availability of classical texts made possible by the invention of printing and the spread of Renaissance humanism to northern universities, traditional views on suicide remained for some time unaffected by the revival of interest in classical philosophy and ethics. To speak, as does Fedden, of "the lag between *elite* and uninformed opinion" ([14], p. 165) is to misrepresent the situation as it existed throughout the sixteenth century. Even Neo-Stoic humanists like the Flemish Justus Lipsius (1547–1606) publicly rejected suicide, counselling instead Stoic imperturbability and Christian patience in the face of evil (but see [34], p. 42, n. 1).

There are two writers whose views on suicide must be examined much more closely as apparent exceptions to the general opposition of sixteenth century writers to suicide: Sir Thomas More and Michel de Montaigne. More (1478–1535), lord chancellor to Henry VIII, was beheaded for his refusal to swear allegiance to the king as supreme head of the English church in place of the pope. A leading humanist, More was a close friend of the painter Hans Holbein and the Flemish humanist Erasmus. In 1515 or 1516, while More was in Flanders on an embassy for Henry, he wrote an account of the island of Utopia (so called from the Greek *ou topos*, 'nowhere'), an imaginary land in the New World. Written in Latin, the work was published at Louvain in 1516 under the supervision of Erasmus, for whose satirical *Praise of Folly*, a work composed in More's house some six years earlier, it was probably intended as a companion piece. Utopia is described as a land that enjoys a community of goods, where its citizens live in harmony and the corruption of private wealth and property is lacking. No one is idle; all work six hours each day and spend their leisure time in intellectual pursuits. There are no wars of aggression and religious toleration is granted to all who recognize the existence of God and the immortality of the soul. One of the many curious features of the Utopians is their practice of voluntary suicide (described in Book 2, chapter 7). The Utopians care for incurables by offering them conversation and relief from pain. However, those who suffer from incurable diseases that are accompanied by continuous pain are urged by the priests and magistrates not to endure their suffering any longer. They are a burden to both themselves and others and should either put an end to life or allow others to release them. In death they will lose nothing but their suffering, and, since the priests interpret God's will, by acting on their advice they will be dying virtuously.

> Those who are moved by these arguments either starve themselves to death of their own accord or through the aid of an opiate die painlessly. If a man is not persuaded to this course, they do not force him to it against his will, nor do they lessen their care of him. To choose death under these circumstances is honorable. But they dishonor a man who takes his own life without the approval of the priests and senate. They consider him unworthy of decent burial and throw his body unburied and disgraced into a ditch (trans. Ogden).

This passage, famous in the history of euthanasia, has elicited much comment. "Here, perhaps for the first time in *Christian* writing," according to Fedden, "is a serious consideration of euthanasia and an unprejudiced approach to a certain type of suicide" ([14], p. 165). "Although the term euthanasia had not yet been coined," writes Fye, "More clearly advocates the active form" ([15], p. 494).

One's view of whether More advocated euthanasia depends to a considerable extent on the opinion one holds (and there are many) of the purpose for which *Utopia* was written. The work is clearly satirical and contains much social criticism of Tudor society, but it is unlikely that More intended it to be a philosophical treatise whose imaginary descriptions were to be taken as serious suggestions for the reform of society. C.S. Lewis argues, convincingly, I think, that its purpose was entertainment: it is "a holiday work, a spontaneous overflow of intellectual high spirits, a revel of debate, paradox, comedy and (above all) of invention, which starts many hares and kills none" ([20], p. 169). There is so much that is amusing, even outrageous, in the customs of the Utopians that it is difficult to sort out that which is to be taken seriously from that which is not. If More did not intend to suggest that we make chamber pots out of gold and silver or that prospective spouses examine their future mates naked to find hidden blemishes, it is not likely that he had in mind to recommend an organized system of euthanasia. It is much easier for More's readers who live in the late twentieth century, when discussion of euthanasia is commonplace, to lift this passage out of its imaginative context and give it more seriousness than its author intended. More was an orthodox Catholic who, as lord chancellor, was severe in his prosecution of heretics and hardly the kind of person to recommend so strongly condemned a practice as suicide. Long after *Utopia* was published, More suggested that it and Erasmus' *Praise of Folly* were books that might better be burned. Perhaps he felt that passages that were meant to entertain were being taken too seriously. In *Praise of Folly* (chapter 31) Erasmus too had commended philosophers and statesmen who killed themselves out of weariness of life as being wiser than old men who sought to cling too long to life. Given the obviously satirical nature of the work, it is difficult to take this statement as an approval of suicide. In fact, Erasmus explicitly scorns it: "This will show you, what would happen if wisdom spread throughout mankind: we'd soon need some more clay and a second Prometheus to model it" (i.e., if everyone practiced suicide the human race would die out).

Michel de Montaigne (1533–92) followed More and Erasmus by a generation. Because of his father's interest in his education, Montaigne imbibed the spirit of the classics from a young age. He was taught to speak Latin before he learned his native French. After studying law and pursuing a public career as a member of the parliament of Bordeaux, he retired to his ancestral chateau in 1571 to pursue a quiet life of reading and reflection surrounded by his classical library of a thousand books. It was in these years that Montaigne invented the essay, both the genre and the term. Until his death in 1592, he wrote and revised his *Essais*, which eventually reached three books. They achieved great popularity and went through several editions in his lifetime.

The tone of Montaigne's essays differs *toto caelo* from even the satirical works of More and Erasmus. The latter were Christian humanists, typical of many of the Northern Renaissance: devoted to the classics and to humanistic studies, they were nevertheless practical and religious. While criticizing the church they remained loyal to its teachings. Montaigne's essays are the product of a very different atmosphere; they reveal an author possessed of a skeptical personality, who is genial, tolerant, non-judgemental, and basically secular in his outlook. His reading of the classics had developed in him essentially a pagan world-view. Sainte-Beuve's bon mot, that he had remained a Catholic long after ceasing to be a Christian, is not unjust. His morality is based on nature and reason rather than revelation and is deeply influenced by both Stoicism (especially that of Seneca) and Epicureanism. Above all he is a Pyrrhonian skeptic, who examines both sides of every question and frequently is satisfied to suspend judgement, convinced that our opinions of religion and morality are usually determined by geography and environment.

After the premature death in 1563 of a boyhood friend, Étienne de la Boétie, Montaigne became preoccupied with the subject of death, a theme of several of his essays. One of the most interesting is 'A Custom of the Isle of Cea' (Book 2, 3), which deals with suicide (*la mort volontaire*). The essay takes its title from an incident recorded in the life of Sextus Pompey, who, while on the island of Cea, was invited to witness the suicide of a 90-year-old woman who chose to end a satisfying life by drinking poison in the presence of her family. The essay is typical of many: Montaigne records numerous examples (most taken from classical sources) of those who, for a variety of reasons, seek to end their lives, together with a number of quotations from Roman writers who are for the most part favorable to the practice of suicide, interspersed with his own comments. "The most voluntary death," he writes, "is the finest. Life depends upon the pleasure of others; death upon our own." The wise man is one who lives not as long as he can, but as long as he should. Nature has given us only one entry into life, but a hundred thousand ways to leave it. When God reduces us to the state in which it is far worse to live than to die, He grants us permission to die. It was Montaigne's practice to juxtapose conflicting opinions in his essays, often without expressing his own view regarding which was preferable. From the number and force of his examples he hoped to draw moral principles. In this vein he admits that the question of suicide is disputed and gives the traditional view:

> For many are of the opinion that we cannot quit this garrison of the world[6] without the express command of Him who has placed us in it; and that it appertains to God who has placed us here, not for ourselves only but for His glory and the service of others, to dismiss us when it shall best please Him, and not for us to depart without His licence: that we are not born for ourselves only, but for our country also, the laws of which require an account from us upon the score of their own interest, and have an action of manslaughter good against us; and if these fail to take cognisance of the fact, we are punished in the other world as deserters of our duty (trans. Cotton).

Several classical quotations are given in support of the traditional view that condemns suicide. Montaigne then returns to discuss "what occasions are sufficient to justify the meditation of self-murder" amongst those who hold that suicide is permitted. He mentions Pliny's belief that there are only three sorts of diseases that permit suicide, the most painful of which is stone in the bladder. The comment is interesting in light of the fact that Montaigne suffered much pain and incapacitation from the stone, for which he repeatedly sought a cure. While both sides of the 'suicide-debate' are presented, Montaigne piles up numerous examples of suicide, probably with the intent of inclining the reader to the belief that the practice can sometimes be justified. Even more important, the whole discussion is taken out of the realm of theology and morals and placed solely in the world of human experience. While opinions may differ, he implies, the experience of so many distinguished men constitutes a presumption in its favor. His conclusion is laconic and (given his own chronic and excruciating pain) poignant: "Pain and the fear of a worse death seem to me the most excusable incitements".

Montaigne was the first significant modern figure to question the Christian position on suicide.[7] For him suicide was no longer grounded in Christian morality (in spite of a formal nod to the Christian position), but was made, as it was in the classical period, a matter of personal decision. Montaigne's *Essais* were much read in the seventeenth century and exercised a good deal of influence on liberal opinion, in England as well as France. But it was in the eighteenth century that his influence was most deeply felt, among his spiritual heirs, the skeptical *philosophes* of the Enlightenment. His view that suicide was not unnatural, since nature herself gave man the ability to die, was taken over by them ([8], p. 60). The immediate influence of Montaigne was, however, much more limited. His real heir intellectually was Pierre Charron (1541–1603), a priest who met Montaigne in Bordeaux, where he became his intimate friend and disciple. Upon his death he administered the last rites of the church and inherited his library. He later became preacher in ordinary to Marguerite of Navarre, wife of Henry IV of France. In 1601 Charron published *De la Sagesse (On Wisdom)*, a popular treatise on moral philosophy in three volumes. The work is indebted to Montaigne, whose ideas it does much to systematize. Based on a sensationalist psychology, the work appeals to reason rather than revelation, divorces morality from religion, and avers that religion is an accident of time and place. Though the book concedes a fideist assent to the truths of Christianity, the tone is skeptical of knowing divine mysteries through reason (hence the need for faith). Like his master Montaigne, he presents both sides of the question whether suicide is permitted, observing that it is not a matter beyond all doubt, as Christian doctrine maintained (*De la Sagesse* 2, 11, 18). Charron observes that suicide has been practiced by "persons eminently great and good of almost every country, and every religion", including Jews and Christians, of whom he cites "those two holy women", Pelagia and Sophronia, who died to preserve their chastity. Suicide is defended, he says, on the ground that if a desire for death is a virtue, the execution of that desire cannot be a sin. The arguments against suicide are, however, "a great deal more substantial and more obligatory". It is condemned not only by Christians and Jews, but by "the Generality of Mankind" and philosophers, and it shows not courage but cowardice. Like Montaigne he asserts that, "considerations of Duty and Religion aside", suicide is not to be attempted "without some very extraordinary and most pressing reason to induce them".

> But indeed there is something to be said (though that something is not enough) for a very urgent and weighty occasion, such as renders Life a perpetual torment, and the thoughts of continuing in it insupportable; such, for instance, as I mentioned formerly, long, acute, extreme pain, or the certain prospect of a very cruel and ignominious death.

In spite of Charron's formal rejection of the practice, he opens the door to the possibility that suicide can sometimes be justified in certain circumstances. Not surprisingly, *De la Sagesse*, with its skeptical point of view, was placed on the Index. In spite of that it became enormously popular in the century after the author's death, going through 35 editions between 1601 and 1672 and surpassing even the popularity of Montaigne's *Essais*. It is incorrect to say, as does Fedden, that "Montaigne's general attitude to suicide we may take to be typical of the enlightened opinion of his time" ([14], p. 163). Montaigne and Charron were the exceptions that proved the rule, insofar as the almost universal disapprobation of suicide in the sixteenth century is concerned. But they were forerunners of a shift in thinking that occurred in the century following, a shift that was in part due to their influence. In arguing for a naturalistic and merely personal basis for suicide, however, they opened a Pandora's box.

EARLY PROTESTANT THOUGHT

The Protestant Reformation, though it divided Christendom and marked a sharp break with much of the doctrine and practice of mediaeval Catholicism, largely took over the Augustinian formulation regarding suicide. There is little specific mention of the subject in the writings of Luther and Calvin, largely because they accepted the orthodox condemnation of the practice. Luther wrote much on the subject of death. It was, he believed, God's punishment for sin in the human race. Because it was penal it was more bitter to Christians than to unbelievers. In fact, he observed, pagans often seem to fear death less than Christians.

Christians could easily bear death if they did not know it as evidence of the wrath of God. This knowledge makes death so bitter for us. But the heathen die in security. They do not see the wrath of God but imagine death to be the end of a man and say: It is merely a matter of one evil little hour. Cicero has pointedly said: *Aut postea nihil erimus aut omnino beati* (After death we shall either be nothing at all, or we shall be altogether blessed). He practically says: Nothing evil can happen to us after death ([22], vol. 1, p. 368).

The difference between Christian and pagan attitudes to death doubtless helps to account for their respective views towards suicide. Protestants were, however, sometimes willing to consider extenuating circumstances. Johannes Neser, a Lutheran pastor at Rothenburg, concerned about the number of suicides in the area in which he lived, delivered three sermons, based on Psalms 77 and 88, which were published in 1613. In these sermons Neser dealt with melancholy and (in the third sermon) with its tendency to drive men to suicide. Citing Luther and other theologians, he asserts that suicide committed by a sane person leads to damnation. But if melancholy or madness leads men to suicide, they are not responsible for their actions and must not be regarded as reprobate. Similarly, in the case of temporary mental imbalance caused by acute pain, we must leave the matter of the soul's eternal destiny in God's hands ([27], p. 276).

John Calvin, like Luther, believed that a Christian might desire death in order to be in God's presence, but he carefully distinguished this from the desire to commit suicide ([30], p. 11, n. 11). Calvin does not discuss suicide when dealing with the sixth commandment (*Institutes* 2.8. 39–40; cp. [30], p. 45). The basis for the commandment forbidding murder, he says, is that every man bears God's image and shares our flesh. "And if we do not wish to renounce all humanity, we ought to cherish his as our own flesh." Hence, his grounds for regarding murder as sin presuppose our desire to protect ourselves from self-destruction. If Calvin did not explicitly derive a condemnation of suicide from the sixth commandment, Anglicans did ([30], pp. 2–3). John King, who later became bishop of London, lecturing on Jonah in 1594, condemned suicide on the basis of an express commandment of Scripture, while citing traditional arguments against it. George Abbot, who was to become Archbishop of Canterbury, took a similar position in his lectures on

Jonah delivered in 1600. Nor did Anglicans lessen the traditional severity of punishment for suicide. Thomas Cranmer's Cathechism was typical: the suicide was "cursed of God and damned forever" ([30], 3–5, 13). Sprott has demonstrated that until the end of the sixteenth century English preachers and moralists all but universally condemned suicides largely in the traditional formulation inherited from Augustine and Aquinas.[8] Far from taking a more lenient stand than traditional teaching permitted, sixteenth-century Anglican divines conceded little by way of exception even to those for whom suicide might be excused by age or insanity.

It is among the English Puritans that we find the greatest development of thinking about suicide among those of Reformed or Calvinistic persuasion. Calvinism entered England by means of Protestants who had fled to Geneva in order to escape the Marian persecution. They returned upon the accession of Elizabeth, bringing with them Calvinistic theology and practices, which they hoped to incorporate into the Church of England. The earliest book published in England on suicide was written by an Anglican clergyman of Puritan sympathies, John Sym's *Lifes Preservative Against Self-Killing* (1637). Sym (1581?–1637) wrote the work as a result of an increase in suicide that he had observed during his own ministry. He wished as well to fill a void that he perceived in the lack of a treatise on the subject. He reiterates the traditional arguments against suicide: its prohibition by the sixth commandment; its defacing of the image of God in man; its unnatural character; and the belief that suicide incurs eternal death. On the other hand, he argues that in the case of disordered minds those who take their own lives cannot be held to be beyond the possibility of redemption. The work is comprehensive and deals with a variety of topics, including the motives to suicide; the kind of behavior that often precedes it; practical measures by which suicide can be averted; and antidotes for those who are melancholic or disposed to self-destruction. Specifically, Sym urges use of the spiritual disciplines, since he regards the best preventive against suicide to be a sanctified life.

Suicide was treated by a number of Puritan writers beginning about 1630. Given the paucity of earlier treatments of the subject, the sudden increase deserves explanation. Some writers have attributed the new interest to a rash of suicides in the mid-seventeenth century (e.g., [30], pp. 32–38). Puritan writers often speak as though the number of suicides had suddenly grown alarming, but it is difficult to determine how far statements of this sort can be pressed statistically. Calvinism itself has been accused of provoking suicide.[9] According to this view the Reformed doctrine of predestination was a 'destructive weapon' ([14], p. 158), whose influence on minds susceptible to melancholy may have induced a fear that they were not among the elect, with the result that some ended their lives in despair. Calvinism is said to have minimized the place of man and deprived him of the feeling of human worth. According to Sprott, "The atmosphere of the Puritan life of piety was heavy with possibilities for suicide" ([30], p. 47). The tendency towards introspection and continual self-examination, as well as the pursuit of holiness and sanctification, were frequently (on this view) productive of a melancholic temperament. Added to these factors, Sprott sees yet another: the belief that God might directly inspire one to take one's own life. Puritans who were subject to inner promptings that they thought revealed God's will might take a temptation to suicide as a direct command from God, which would override perceived biblical injunctions against it.

Though there seems much that is plausible in the view that Calvinism encouraged suicide, there is little evidence to support it. Calvin had taught, and the Puritans agreed, that evidence of election was to be found in personal righteousness and good works (Calvin, *Institutes* 3.23.12). Speculation about predestination was discouraged (*Institutes* 3.21.1). Tender consciences might be led into an excess of introspection, but Calvin and the Puritans warned that self-examination could be carried too far (see *Institutes* 3.24.4). The doctrine of predestination was not meant to produce fear and disquiet but confidence and hope: "In order to free us of all fear and render us victorious amid so many dangers, snares, and mortal struggles, he promises us that whatever the Father has entrusted into his keeping will be safe" (*Institutes* 3.21.1). According to the Thirty-Nine Articles of the Church of England, the doctrine of predestination is "full of sweet, pleasant and unspeakable comfort to godly persons" (Article XVII). "When Judas hanged himself," writes C.S. Lewis, "he had not been reading Calvin" ([20], p. 34, n. 1). And while Puritans sought divine guidance through dreams, promptings, and the like, they were careful to do so only within the prescribed limits of Scripture. Sprott is incorrect to say that "in Puritan experience the authority of Scripture did not clearly prescribe suicide" ([30], p. 52). To Sym and other Puritan writers suicide was "utterly unlawful" because it was forbidden by God's law as revealed in Scripture. Puritan writers charged Quakers and other enthusiasts with divorcing the "inner light" from the objective revelation of Scripture. As Sym asserts, a temptation to kill oneself could not possibly come from God and must therefore proceed from Satan.

How then do we account for the frequent discussion of suicide by Puritans in the seventeenth century? A partial explanation is that suicide was 'in the air', a subject of conversation and growing debate. More than this, however, was the Puritan belief that suicide was one of Satan's snares, which particularly accompanied melancholy. Sermons, casuistical treatises, and books of pastoral theology and spiritual discipline dealt with it, as they dealt with many other temptations. Several works discuss it in their advice on how to treat melancholy. If Christians drawn to suicide were to be saved from this act, they must be shown that it was a sin whose eternal consequences were terrifying, and they must be given practical advice regarding how to avoid it. It was precisely this that Puritan writers on the subject attempted to do.

Nearly a century ago Emile Durkheim pointed out that there was a correlation between religious beliefs and the proportion of suicides in several parts of Europe ([11], pp. 152–170). Durkheim found that the proportion of suicides in Catholic countries or provinces was significantly lower than in Protestant ones, the difference varying between 20 to 30 percent and 300 percent. Durkheim observed that both confessional systems equally condemned suicide. What then accounted for the marked difference between Protestant and Catholic countries? His answer was that Protestantism permitted a far greater degree of free enquiry, while Catholicism by its authoritarian system obtained much more uniformity of belief. Protestantism permitted greater freedom to the individual to define his own beliefs; as a result, it produced a less integrated community than that found in Catholicism. Similarly, Sprott has adduced evidence to show that "from the mid-1640's to 1660, suicide was some fifty percent commoner than in the years immediately before or after" ([30], pp. 32–33). He speaks of this as the "Puritan epidemic of suicide", and, while admitting that the causes were many and complex, he considers Puritan beliefs (particularly the Puritans' reliance on inner prompting in seeking God's will) a contributing factor ([30], pp. 47–53). Were there characteristics peculiar to Protestantism that might in some way, even indirectly, have been conducive to the encouragement of suicide?

The higher proportion of suicides in traditionally Protestant areas of Europe probably finds an explanation less in theological differences than in divergent approaches of Protestants and Catholics to moral theology. The Roman Catholic Church over time had fashioned an impressive moral position that was uncompromising in its demands, buttressed by an authoritarian hierarchy, and maintained by an efficient system of ecclesiastical discipline. The foundation of its moral stance was the tradition of the church. By contrast the ground of authority for Protestantism was, from the time of the Reformation, *sola scriptura*. While Protestants were willing to take into account tradition and reason, their procedure in determining matters of faith and morals was to collect and exegete the relevant texts of Scripture in order to establish normative teaching. There were moral issues, however, that did not permit ready solutions because the biblical teaching was not clear. In such instances Protestants either drew inferences from biblical principles or from passing references in Scripture. Thus, while they usually accepted traditional Christian moral positions, they often lacked unanimity of opinion at least in details. This was a reflection of disagreements over the exegesis or understanding of Scripture. Even when theologians agreed on a point of doctrine or moral theology, their opinion did not constitute a magisterium. In dealing with suicide, Protestant theologians struggled with the scriptural examples of Samson, who had killed himself along with the Philistines (Judges 16:21–30, esp. v. 30), and Jonah, who asked that he be cast into the sea to calm the storm (Jonah 1:11–12) ([30], 3–5, 51–52). Protestants, too, were willing to remain silent where Scripture did. This meant that they were more inclined than Catholics to admit exceptions and to entrust doubtful cases to the judgement and mercy of God: hence their willingness to believe that suicides suffering from mental disorders might be saved. "In practice", writes Sprott, "Anglicans from John Foxe to Jeremy Taylor seem regularly to have held that though the deed was damnable, the judgement whether a pious and distressed self-murderer were lost might in charity be left to God" ([30], p. 13). There were, of course, numerous differences in degree between Anglicans and Puritans, but there was broad agreement in their approach to many issues of moral theology. Hence, while the Catholic position on suicide remained unbending and its condemnation of the act left little hope for mercy, Protestants, although maintaining that suicide was a terrible sin, took a somewhat more flexible position. The difference is nicely illustrated by a discussion in 1600 between the Anglican Dr. George Abbot, future Archbishop of Canterbury, and a Catholic Jesuit, John Gerard, regarding whether suicide was an unforgivable sin.

> "But", said the doctor, "we don't know whether this was such a sin".
>
> "Pardon me", I [Gerard] said, "it is not a case here of our judgement. It is a question of God's judgement; He forbids us under pain of hell to kill anyone, and particularly ourselves, for charity begins at home".
>
> The good doctor was caught. He said nothing more on the point.[10]

Even if the Protestant position provided slightly more encouragement to some who were contemplating suicide, it is, however, insufficient to account fully either for the apparent increase in English suicides in the mid-seventeenth century or for the relative proportion of suicides in Protestant and Catholic areas of Europe at a later period.[11]

THE SEVENTEENTH CENTURY

The end of the sixteenth century witnessed a growing interest in the subject of suicide. Previously discussed most often in casuistical works or peripherally as it related to more central themes in moral theology, it came increasingly into its own as a subject for extended treatment, both in academic dissertations written in continental universities and in treatises intended for private or public circulation ([30], pp. 20, 25–26). Sprott attributes this new interest to the crises occasioned by religious persecution and war ([30], pp. 16–17). A more likely explanation is to be found perhaps in the influence of classical authors (especially Stoic philosophers like Seneca), as well as in the influence of Montaigne and Charron, whose works were translated into English in the early years of the seventeenth century. The impact of their skeptical thought can be seen, for example, in the views of the Abbot of St.-Cyran, father of the Jansenist movement in French Catholicism, who defended suicide in his *Casus regius*, written in 1609. St.-Cyran argued that, though suicide is forbidden by the sixth commandment, circumstances sometimes arise that justify it even as they sometimes justify homicide. In order to support his position he adduced some 34 instances in which, he asserted, the act would not be immoral. His *Casus regius* enjoyed great popularity in France, while arousing much opposition. A year or two before its publication, the English poet and divine John Donne (1572–1631) wrote his famous *Biathanatos*, the first defense of suicide written in English, probably in 1607 or 1608. Donne took his title from a Greek word, *biathanatos*, popular with Renaissance writers, that denoted a violent death and (by extension) a self-inflicted one ([9], pp. 402–403). The work was not intended for publication, but was circulated privately among friends. After Donne took Anglican orders in 1615 he apparently felt uncomfortable about the work, as he indicated in 1619 in a letter to his friend Sir Robert Ker: "Let any that your discretion admits to the sight of it, know the date of it; and that it is a Book written by *Jack Donne*, and not by D[octor]. *Donne*" ([2], p. 201). Donne told Ker that he had "always gone so near suppressing it as that it is only not burnt", and he advised him to "keep it...with the same jealousy...Publish it not, yet burn it not" ([10], p. xv). The book was finally published between 1644 and 1647, after Donne's death, by his son John

Donne the younger, who published a number of his father's works posthumously. Donne always displayed some ambivalence about the work. His refusal to publish it probably stems from his reluctance to expose himself to criticism for having challenged the traditional Christian view. Donne recognized that his book dealt with "a misinterpretable subject". On the other hand, in a letter to his friend Edward Herbert, to whom he gave a copy, he wrote that if he were to destroy the manuscript "those reasons by which that act should be defended or excused were also lost with it" ([10], p. xv). There is little indication whether Donne ever abandoned the views he set forth in *Biathanatos*. As Dean of St. Paul's he would doubtless be reluctant to have his name indentified with so controversial a volume: hence the distinction that he makes between "Jack Donne" and "D[r]. Donne" perhaps reflects "a difference in situation and stance rather than a difference in opinion" ([10], p. xvi).

Donne begins his work with a reference to his personal situation as regards the subject with which he intends to deal:

> I have often such a sickly inclination; and whether it be because I had my first breeding and conversation with men of a suppressed and afflicted religion, accustomed to the despite of death and hungry of an imagined martyrdom, or that the common enemy find that door worst locked against him in me, or that there be a perplexity and flexibility in the doctrine itself, or because my conscience ever assures me that no rebellious grudging at God's gifts, nor other sinful concurrence, accompanies these thoughts in me, or that a brave scorn, or that a faint cowardliness beget it, whensoever any affliction assails me, methinks I have the keys of my prison in mine own hand, and no remedy presents itself so soon to my heart as mine own sword.

The attraction that suicide had for Donne may have grown out of the spells of melancholy with which he was afflicted, a condition that was doubtless exacerbated by his poor health. There is frequent reference in his letters to both. The book was written during a period of intense mental conflict for Donne. Perhaps Bald is correct in seeing *Biathanatos* as the product of Donne's melancholic moods, which he wrote "to overcome a temptation, not by trying to banish it altogether from the mind, but by giving it full place there and at the same time rendering it innocuous by transferring it from the place of action to that of learned investigation and contemplation" ([2], p. 158). Learned the book certainly is. Donne cites some 200 authors and sources (which he lists at the beginning of the work), classical, patristic, mediaeval, and contemporary. He displays a thorough acquaintance with civil and canon law, of which he had made a special study. In spite of the author's reluctance to publish *Biathanatos*, it bears all the marks of a formal treatise, carefully arranged, heavily annotated, and giving every indication of being intended for a wide reading public.

Biathanatos is a casuistic treatise modelled on similar works that were produced by moral theologians to deal with cases of conscience. On the title page Donne describes it as "A Declaration of that Paradox or Thesis, that Self-Homicide is not so naturally Sin that it may never be otherwise". It is a paradox (in his view) in the sense that his position is contrary to the accepted opinion regarding suicide. Donne turns all the traditional arguments against suicide on their head, though in doing so he employs a certain amount of special pleading. He does not cite Montaigne or Charron, and, although there is a skeptical strain in the first two parts of the book, Donne appears to be writing less from a skeptical than from a Christian point of view, unless we take the work to be written in the mocking manner of a literary paradox.[12]

Donne's definition is broader than modern definitions of suicide and includes all cases of voluntary death. He does not suggest that in all cases suicide is justified, merely that in some instances it may be, and the act is not the 'irremissible' sin that traditional Christian teaching has made it to be. He attempts to show that it does not violate the law of nature (which requires self-preservation), the law of reason (as found in civil and canon law), or the law of God (as revealed in Scripture). He argues that there is no law that does not admit of exceptions. The sin that attaches to an act does not lie in the act itself but in the fact that it is forbidden by God. God sometimes commands an act that is ordinarily regarded as sinful. Therefore the rightness of an act lies in the circumstances. There are exceptional cases that require individual judgement. Donne points out that the sanctions of civil and canon law against suicide are based on the desire to deter the practice. The severity of English law simply reflects the fact that human beings in general possess a natural urge to die, not in order to achieve annihilation, but to attain a spiritual afterlife free from the confines of the body. Canon law has never regarded suicide as wrong. While it has been condemned by church councils, Donne draws a distinction between the decrees of councils and canon law. The latter is binding, while the former are not. In typical Renaissance fashion Donne (like Montaigne) cites many examples of famous suicides from the classical world. But he names as well Christians who were martyrs. For Donne martyrdom was a form of "self-homicide". It was perhaps due to the influence of his Catholic upbringing that he always considered martyrdom to be the highest role that a Christian could attain. Donne adduces not only the Christian martyrs in support of his position but even Christ himself, who "chose that way for our redemption to sacrifice His life and profuse His blood". He concludes by stating that, while suicide undertaken for self-interest is not permitted, those suicides that imitate Christ's self-sacrifice and are undertaken for the glory of God are. It falls to the conscience to determine whether the act in any particular situation is justifiable.

The relationship of melancholy to suicide was a subject of much interest in the sixteenth and seventeenth centuries, as may be seen from the frequency with which the subject was discussed.[13] Melancholy had often enough been described by Christian writers in a pastoral context, but it began to attract broader attention, particularly in Elizabethan England. The Renaissance in Europe gave rise to a widespread sense of melancholy, characterized by feelings of sorrow, nostalgia, the realization that life is brief and its pleasures fleeting, and a general *taedium vitae*, which had no apparent external cause. This malaise was the subject of art (e.g., Duerer's 'Melencolia') and literature (e.g., Shakespeare's tragedies). Shakespeare presents some fourteen suicides in eight of his tragedies, of which *Hamlet*, with its melancholic hero who debates suicide in the soliloquies, is the best known. Aristotle remarks that philosophers and lawgivers tend to have melancholic temperaments, and the malady (whether real or affected) seems to have been common among Renaissance men of letters. It is against this background that the greatest work on the subject, Burton's *Anatomy of Melancholy*, was written. Robert Burton (1577–1640) came to Oxford as a student, later becoming, as well as Vicar of St. Thomas, a tutor and librarian, and he remained at Oxford till his death. He is said to have predicted the approximate date of his death from astrological calculations, and it was rumored (without any evidence) that he committed suicide in order to bring the prophecy to pass. In 1621 he published the book that made his fame, *Anatomy of Melancholy*, under the name of Democritus Junior. The work went through six editions, each with alterations and revisions, the last appearing in 1651. This highly original book is intended to deal with melancholy as a medical and psychological phenomenon, but it contains many digressions together with all sorts of curious lore and classical allusions and quotations. It is divided into three parts, the first dealing with the innumerable causes and symptoms of melancholy, the second with its cure, and the third with the various forms of melancholy that arise from love, jealousy, and religion. Burton declares in his preface that he wrote the book to escape melancholy himself. He attributes the affliction to an excess of black bile; he also mentions such natural causes as the influence of the stars, education, heredity, diet, and environmental factors. He ends the first part of his work with a discussion of suicide. For Burton suicide is not an act that forever results in eternal damnation. Rather, it is the result of melancholy that desires self-destruction:

In other diseases there is some hope likely, but these unhappy men are born to misery, past all hope of recovery, incurably sick, the longer they live the worse they are, and death alone must ease them ([4], p. 285).

Burton cites numerous examples of classical suicide and observes that the act has been recommended by many famous persons; at the same time he rejects the pagan or Stoic rationale for suicide. However, he argues that there may be circumstances that mitigate the guilt of suicide, particularly in the case of those who are suffering from madness or melancholy, for "in extremity, they know not what they do, deprived of reason, judgement, all, as in a ship that is void of a Pilot, must needs impinge upon the next rock or sands, and suffer shipwreck" ([4], p. 287). About their eternal destiny he writes:

Of their goods and bodies we can dispose; but what shall become of their souls, God alone can tell; his mercy may come *inter Pontem et Fontem, inter Gladium et Jugulum*, betwixt the bridge and the brook, the knife and the throat.... We ought not to be so rash and rigorous in our censures, as some are; charity will judge and hope the best; God be merciful unto us all ([4], p. 288).

In his willingness to accept mitigating circumstances and to trust in God's mercy for a suicide who is *non compos mentis*, Burton reflects what had become the standard Protestant view.

By the mid-seventeenth century the subject of suicide was attracting a good deal of attention, stimulated in part by the publication of Donne's *Biathanatos* in the 1640's. In 1653 a rejoinder was published in the form of a poem in twelve cantos by Sir William Denny, entitled *Pelecanicidium: Or the Christian Adviser against Self-Murder*. The work takes its title from the pelican, which was said to sacrifice itself for its offspring by piercing its breast to feed them with its own blood. Denny wrote the work, he tells us, after hearing of so many suicides, "lest the Frequency of such Actions might in time arrogate a Kind of Legitimation by Custom, or plead Authority from some late publisht Paradoxes, That Self-homicide was Lawfull" ([30], pp. 31–32). Doubtless his reference is to *Biathanatos*. Whether or not it is true that its publication precipitated a large number of suicides ([9], p. 421), *Biathanatos* provoked much debate. In a disputation held at Cambridge between 1655 and 1661, the Regius Professor of Divinity, Anthony Tuckney, analyzed Donne's thesis in detail and presented a closely-reasoned argument against suicide. At about the same time the question of whether suicide was lawful was debated at Oxford as well in a divinity disputation at Queen's College ([30], pp. 62–66). Clearly, the matter was one of widespread interest, and the opponents of suicide were anxious to counter the arguments in its favor. The subject continued to be treated in Anglican casuistical works, such as Jeremy Taylor's *Doctor Dubitantium, or the Rule of Conscience*, published in 1660. Taylor employs the usual arguments against suicide: it is contrary to nature; it displays cowardice; it contravenes God's law. The traditional position, with little change or variation, continued to be iterated in published sermons into the eighteenth century ([30], pp. 100–102).

The Christian condemnation of suicide influenced even so independent a thinker as John Locke (1632–1704), political philosopher and founder of English empiricism. In his *Second Treatise of Civil Government* (1690) Locke undertook a defense of the Glorious Revolution of 1688, which had established the supremacy of Parliament, by employing the social-contract theory of government developed earlier by Thomas Hobbes. Like Hobbes and the school of natural-law theorists, Locke argues that at one time all men lived in a state of nature in which there was no government but there were complete freedom and equality. The only law that existed was the law of nature, which each person maintained for himself in order to guard his natural rights to life, liberty, and property. Since the inconveniences of life in the state of nature were greater than its advantages, men agreed to establish a civil society and government. But they did not grant to it absolute power: the state may not, says Locke, violate natural rights, which it is bound to protect. Even life in the state of nature has restrictions, one of which limits one's right to take one's own life.

> But though this be a state of liberty, yet it is not a state of licence; though man in that state have an uncontrollable liberty to dispose of his person or possessions, yet he has not liberty to destroy himself, or so much as any creature in his possession, but where some nobler use than its bare preservation calls for it (Ch. 2, 6).

The law of nature teaches mankind that "no one ought to harm another in his life, health, liberty or possessions", for men are the creation and servants of God, placed in the world to do His business and "made to last during His, not one another's pleasure". Just as everyone is "bound to preserve himself, and not to quit his station wilfully", so he ought to do whatever he can to preserve the life, liberty, and property of another. Locke's condemnation of suicide on theological grounds (man is God's property and workmanship) is notable chiefly because it is inserted in a work that otherwise relies on non-theological reasoning. Perhaps Locke owes his conviction on the matter to his Puritan background. Locke's family were Puritan (his father had fought on the Parliamentary side in the Civil War), and as he attended Oxford in the

1650's, when it was under the domination of the Puritans, he may have been influenced by the debates there following the publication of *Biathanatos*. In his *First Treatise of Civil Government* (Ch. 1, 86) Locke discusses the "natural Inclination" of man to preserve himself, a desire that is shared by the animal kingdom. There is an even greater desire, however, that sometimes overrules the instinct to preserve oneself, and that is the wish to protect offspring, for God

> makes the Individuals act so strongly to this end that they *sometimes* neglect their own private good for it, and *seem* to forget that general Rule which Nature teaches all things of self-Preservation, and the Preservation of their Young, as the strongest Principle in them over rules the Constitution of their particular Natures (*First Treatise*, Ch. 1, 56).

Locke may have had in mind the preservation of one's offspring when he speaks in the *Second Treatise* of "some nobler use" that might justify self-destruction.

It has been argued from passages in his *Essay Concerning Human Understanding* (Book 2, Ch. 21, 54 and 59) that Locke considered suicide a "normal" response of those suffering great pain or "violent passion" ([35], p. 172). In these passages, which do not explicitly refer to suicide, Locke merely acknowledges, however, the weakness of human nature when faced with overwhelming suffering or desire, reminds the reader of God's mercy in recognizing his weakness, and urges him to control his passions (54). In a similar vein he cautions: "*Necessitas cogit ad turpia*; and therefore there is great reason for us to pray, 'Lead us not into temptation'" (59). There is, in fact, nothing in the *Essay* regarding suicide that indicates Locke's approval of the practice. If we can infer anything from these passages, it is that Locke, like many other Protestants, while condemning suicide, is willing to entrust its victims to God's mercy.

> God, who knows our frailty, pities our weakness, and requires of us no more than we are able to do, and sees what was and what was not in our power, will judge as a kind and merciful Father (54).

Another passage that has been cited to support the view that Locke was not absolutely opposed to suicide is found in the *Second Treatise* in a discussion of slavery (Ch. 4, 22–23). Locke writes:

> For whenever he find the hardship of his slavery [to] outweigh the value of his life, it is in his power, by resisting the will of his master, to draw on himself the death he desires (22).

A few lines earlier, however, he explicitly states that man does not have power over his own life and therefore cannot, by enslaving himself, grant to another that which he does not himself possess. Locke merely applies to the condition of slavery what he observes elsewhere: that pain and suffering cause men to take extreme measures that, while not justified, are understandable. It is difficult to see in Locke any real advance in the acceptance of suicide over the traditional English Protestant position. There is no indication that his view in the *Essay Concerning Human Understanding*, based on the law of nature, is at variance with his theologically-based rejection of suicide in the *Second Treatise*. In the latter Locke appeals for support to special revelation, in the former to the law of nature. He takes them as complementary sources of knowledge and fully compatible with one another. Both teach that life is an inalienable right that can be neither taken nor given away.

If the publication of Donne's *Biathanatos* provoked a vigorous defense of the traditional Protestant position, it also attracted a small but significant number of supporters to its thesis. The growing rationalism of the latter half of the seventeenth century provided fertile soil for challenges to the traditional condemnation of suicide. The tendency to bring every question before the bar of human reason and to seek to fit all knowledge within the framework of natural law invariably weakened the theological foundations of philosophy and morals and led to the rise of deism. The origins of deism can be traced to Lord Herbert of Cherbury, whose publication of *De Veritate* (Paris: 1624; London: 1633) marked him as the Father of English deism. The deists rejected the concepts of special revelation and a supernatural basis of religion, seeking instead to construct a natural religion on the universal belief in God and immortality. Sprott calls the period from 1683 to 1720 the "libertine era of suicide", when a number of voices, reflecting the growth of rationalism, were raised in defense of suicide. One of the most prominent was Charles Blount (1654–1693), a disciple of Lord Herbert, who was active in undermining the credibility of Christianity, insisting that whatever was contrary to human reason was "absurd, and should be rejected". Blount was a prolific writer of books that attacked Christian dogma, organized religion, belief in miracles, and the biblical narratives. As early as 1680 he spoke approvingly of Donne's *Biathanatos*. In 1693 he ended his own life because he could not by law marry the sister of his deceased wife. By his own suicide, Blount, the best known deist of the day, pointed to a connection between the skepticism of his philosophy and espousal of suicide. This connection was made clear by Charles Gildon's defense of Blount's suicide in his 'Account of the Life and Death of the Author', which he included in an edition of

Blount's *Miscellaneous Works* that was published posthumously in 1695. Gildon argues that the law of self-preservation admits of limitations and exceptions, and that suicide might be justified when required by honor, virtue, honesty, or the public good. The principle of self-preservation no longer obtains when life ceases to be "a Good". It is for each individual to determine his destiny. "Every man is *sui Juris*, that is Judge, or rather disposer of himself" ([30], pp. 72–73).[14] Not surprisingly, deists were soon being accused of teaching that suicide was a principle that was in harmony with nature and reason, although a number of leading deists refrained from doing so. Some, however, may have regarded the practice as too controversial to be espoused openly, especially since they categorically denied advocating libertinism in morals ([30], pp. 73–78).

Epicureanism, which was revived by the French priest, philosopher, mathematician, and scientist Pierre Gassendi (1592–1655), came to enjoy popularity with a number of English deists after his works began to be translated into English in 1660. Epicurus and Lucretius were widely read and several translations of their works (including a translation of Lucretius by Dryden) appeared. Epicurus' advocacy of suicide attracted attention, but even some of those who opposed his position on Christian grounds undertook to demonstrate that it was not necessarily contrary to the law of nature.[15]

According to Sprott, the incidence of suicide in Greater London in the last two decades of the seventeenth century rose by fifty per cent. It increased again in the first half-dozen years of the eighteenth century ([30], p. 71). To what extent this rise was influenced by the rationalistic defense of the practice is, of course, uncertain. It is worth noting, however, that in 1700 there was sufficient interest to call for a second edition of *Biathanatos*, which by then had become well known. Suicide also became a theme in Restoration drama. The subject had already been treated in the French theatre in the plays of Corneille and Racine, and it now appeared in some of the best-known English plays of the late seventeenth century, including Thomas Otway's *Venice Preserv'd* (1682) and John Dryden's *All for Love* (1678), the latter a retelling of the story of Antony and Cleopatra. In both suicide is depicted sympathetically, reflecting a new tolerance of, perhaps even admiration for, heroic suicide and the suicide of lovers.

If the influence of rationalism led some Englishmen to challenge the traditional condemnation of suicide, attitudes in France remained conservative throughout the seventeenth century. While French tragedies that dealt with the classical period sometimes depicted suicide as a noble sacrifice, philosophers and theologians continued to condemn it. There was a strong skeptical undercurrent in France in the seventeenth century, but it did not lead to the public espousal of suicide. "Malebranche, Nicole, Arnaud, Descartes, La Mothe le Vayer, and theologians of every stripe, Jansenist, Jesuit and Protestant, were in agreement" ([8], p. 50). Legal and ecclesiastical approaches retained a basically mediaeval outlook. In 1670 Louis XIV promulgated the *Ordonnance criminelle*, which made suicide a major crime, like heresy and treason, and extended to all suicides the penalty formerly reserved only for criminals who took their own lives. Their names were to be perpetually disgraced, and in each case the corpse was to be dragged through the streets face down and hanged or cast on the town dump. Although this reactionary legislation was never completely enforced, it provoked much opposition in the next century, particularly from philosophers and reformers. It was in the early eighteenth century, too, that the French and other continental Europeans began to speak of melancholy as the "English disease". The English were widely believed to be willing, as a result of their melancholic disposition, to put an end to their lives with little or no reason. Their propensity for suicide was variously explained as the result of environmental factors (chiefly diet and climate), religion, inactivity, and even failure to marry.[16] The French were sufficiently interested in this "English" phenomenon to borrow the word "suicide", which had been coined in England in the mid-seventeenth century. Daube (following the *Oxford English Dictionary*) traces the origin of the word to Walter Charleton's translation of Petronius' *Satyricon*, which appeared in 1651 ([9], pp. 421–22). Alvarez, however, has discovered in Sir Thomas Browne's *Religio Medici*, which was written in 1635 and published in 1642, an earlier mention of the word:

> Yet herein are they in extreams, that can allow a man to be his own Assassine, and so highly extol the end and suicide of Cato. This is indeed not to fear death, but yet to be afraid of life (Part 1, Sect. 44).

The word caught on slowly in English: it did not appear in Samuel Johnson's *Dictionary* (1755). But it gradually came to replace other phrases, such as "self-murder", "self-destruction", "self-homicide", and the like ([1], p. 12). It was taken over into French by the Abbe Pierre Desfontaines in the eighteenth century, achieved widespread popularity because of Voltaire's adoption of the word, and even gave rise to a late Latin form, *suicidium*.

CONCLUSION

The sixteenth and seventeenth centuries form an important transitional period in the development of the ethics of suicide. The period began with the Augustinian position uncontested. The early Renaissance did not challenge that position, although over time the writings of classical philosophers who defended suicide came to influence thinking on the subject. Montaigne and Charron marked the first genuine departure from the Augustinian view. They represented a skeptical and fideist point of view, but their challenge to the orthodox condemnation of suicide was tentative and cautious; and although their writings remained influential throughout the seventeenth century, their greatest impact was not felt until the Enlightenment. The Protestant Reformation saw a different departure from the Augustinian formulation. While Protestants condemned suicide quite as strongly as Catholics, their emphasis was not the same. Whereas Catholic moral theology left no hope for the eternal destiny of a suicide, many Protestants were unwilling to deny absolutely the possibility of repentence and God's mercy. Hence the Protestant position was often (but not invariably) less rigid. Protestants did not wish to encourage a more tolerant attitude to suicide, but the inevitable result was (here as elsewhere) a greater liberty of thought and hence freer discussion. Perhaps it was her religious development that placed England in the forefront of the discussion of suicide. Far-reaching in its influence was Donne's *Biathanatos*, which, while not the result of a rationalist or skeptical mind, nevertheless proved attractive to those rationalists who later found support in Donne's learned work for their defense of suicide. It must be emphasized that the number of those who attacked the orthodox position that suicide was a sin and not a matter of personal choice was small throughout the seventeenth century. Their importance lies in the fact that they formulated a non-Christian position that divorced the morality of suicide from theological considerations. Although this position remained very much a minority view as long as Christian presuppositions remained dominant, it was adopted and propounded with great success by the *philosophes* in the eighteenth century, when the rationalism of the Enlightenment provided a more receptive climate.[17, 18]

NOTES

[1] "Thou shalt not kill" (Ex. 20:13; Dt. 5:17). This commandment is numbered sixth by the Greek Orthodox and most Protestant Churches and fifth by the Lutheran and

Roman Catholic Churches.

[2] I use the masculine pronoun generically in order to avoid the unnecessarily awkward circumlocutions that non-gender usage so often requires.

[3] See [27], p. 273; and [14], pp. 27–48; but cp. [16], p. 333.

[4] See Kristeller, 'The Moral Teaching of Renaissance Humanism', in [19], pp. 20–68.

[5] See [14], pp. 160–166; and [17], pp. 90–94.

[6] For examples of this commonplace, frequently used from antiquity through the eighteenth century in arguing against suicide, see [18], p. 273, n. 3.

[7] See [26], p. 377; [8], p. 60; [33], p. 43.

[8] See [30], pp. 1–20; *pace* [13], p. 31–32 (but see in reply [36]).

[9] See [14], pp. 157–158; [26], p. 377; [30], pp. 47–54.

[10] This incident is quoted in [31], p. 153.

[11] Perhaps a more likely, if partial, explanation of the latter phenomenon lies in the growth of what H.R. Trevor-Roper calls the "tolerant, sceptical rationalism" of the intellectual heirs of Erasmus that gradually replaced orthodox Protestantism in the Protestant countries of Europe and is likely to have been less opposed to suicide than orthodox Protestantism. See his 'The Religious Origins of the Enlightenment' in [32], pp. 193–236, esp. p. 223.

[12] For a brief summary of this view, see [7], p. 80, with the bibliography cited, especially the works of Colie, Malloch, and Webber. Against the "paradox" school of interpretation see [10], pp. xx–xxi. For another view see [34].

[13] See [14], pp. 166–185; and [26], pp. 378–379.

[14] Ten years later Gildon retracted this view and formally condemned suicide ([30], pp. 90–91).

[15] See, e.g., the arguments of Walter Charleton quoted in [30], pp. 79–80.

[16] See [27], pp. 278–280; and [23], pp. 428–429.

[17] See [8]; and [23], pp. 409–437.

[18] This publication was supported in part by a College of Liberal Arts Research Grant from Oregon State University. I thank Drs. Darrel W. Amundsen and Lisa T. Sarasohn for their helpful suggestions.

BIBLIOGRAPHY

[1] Alvarez, A.: 1980, 'The Background', in M. Pabst Battin and David J. Mayo (eds.), *Suicide: The Philosophical Issues*, St. Martin's Press, New York.

[2] Bald, R.C.: 1970, *John Donne: A Life*, Oxford University Press, New York and Oxford.

[3] Browne, Sir Thomas: 1923 (1881), *Religio Medici*, W.A. Greenhill (ed.), Macmillan, London.

[4] Burton, Robert: 1927, *The Anatomy of Melancholy*, Chatto & Windus, London.

[5] Calvin, J.: 1960, *Institutes of the Christian Religion*, trans. by F.L. Battles, John T. McNeill (ed.), 2 vols., Westminster Press, Philadelphia.

[6] Charron, Pierre: 1729, *Of Wisdom*, trans. by G. Stanhope, 3rd ed., vol. 2, Wilkin, Bonwicke, *et al.*, London.

[7] Colie, R.L.: 1973, 'Literary Paradox', in Philip P. Wiener (ed.), *Dictionary of the History of Ideas*, vol. 3, Charles Scribner's Sons, New York, pp. 76–81.

[8] Crocker, L.G.: 1952, 'The Discussion of Suicide in the Eighteenth Century', *Journal of the History of Ideas* 13, 47–72.
[9] Daube, D.: 1972, 'The Linguistics of Suicide', *Philosophy and Public Affairs* 1, 387–437.
[10] Donne, John: 1982, *Biathanatos*, M. Rudick and M.P. Battin (eds.), Garland Publishing Co., New York.
[11] Durkheim, E.: 1951 (1897), *Suicide: A Study in Sociology*, trans. by J.A. Spaulding and G. Simpson, The Free Press, Glencoe, Illinois.
[12] Erasmus, D.: 1971, *Praise of Folly*, trans. by B. Radice, Penguin Books, Harmondsworth, England.
[13] Faber, M.D.: 1967, 'Shakespeare's Suicides: Some Historic, Dramatic and Psychological Reflections', in E.S. Shneidman (ed.), *Essays in Self-Destruction*, Science House, New York, pp. 30–58.
[14] Fedden, H.R.: 1938 (rpt. 1972), *Suicide: A Social and Historical Study*, Benjamin Blom, New York.
[15] Fye, W.B.: 1978, 'Active Euthanasia: An Historical Survey of its Conceptual Origins and Introduction into Medical Thought', *Bulletin of the History of Medicine* 52, 492–502.
[16] Graziani, R.: 1969, 'Non-Utopian Euthanasia: An Italian Report, c. 1554', *Renaissance Quarterly* 22, 329–333.
[17] Gruman, G.J.: 1973, 'An Historical Introduction to Ideas about Voluntary Euthanasia: With a Bibliographic Survey and Guide for Interdisciplinary Studies', *Omega* 4, 87–138.
[18] Hirzel, R.: 1908, 'Der Selbstmord', *Archiv für Religionswissenschaft* 11, 75–104, 243–284, and 417–476.
[19] Kristeller, P.O.: 1965, *Renaissance Thought II: Papers on Humanism and the Arts*, Harper & Row, New York.
[20] Lewis, C.S.: 1954, *English Literature in the Sixteenth Century Excluding Drama*, Clarendon Press, Oxford.
[21] Locke, John: 1952, *Locke, Berkeley, Hume*, Great Books of the Western World, vol. 35, Encyclopedia Britannica, Chicago.
[22] Luther, M.: 1959, *What Luther Says: An Anthology*, Ewald M. Plass (ed.), 2 vols., Concordia, St. Louis, Missouri.
[23] McManners, J.: 1981, *Death and the Enlightenment: Changing Attitudes to Death among Christians and Unbelievers in Eighteenth-century France*, Clarendon Press, Oxford.
[24] Montaigne, Michel Eyquem de: 1952, *The Essays*, trans. by Charles Cotton, W. Carew Hazlitt (ed.), Great Books of the Western World, vol. 25, Encyclopedia Britannica, Chicago.
[25] More, Sir Thomas: 1949, *Utopia*, trans. by H.V.S. Ogden, Appleton-Century-Crofts, New York.
[26] Noon, G.: 1978, 'On Suicide', *Journal of the History of Ideas* 39, 371–386.
[27] Rosen, G.: 1971, 'History in the Study of Suicide', *Psychological Medicine* 1, 267–285.
[28] Sacharoff, M.: 1972, 'Suicide and Brutus' Philosophy in *Julius Caesar*', *Journal of the History of Ideas* 33, 115–122.
[29] Seidler, M.J.: 1983, 'Kant and the Stoics on Suicide', *Journal of the History of*

Ideas 44, 429–453.
[30] Sprott, S.E.: 1961, *The English Debate on Suicide from Donne to Hume*, Open Court, LaSalle, Illinois.
[31] Sprunger, K.L.: 1972, *The Learned Doctor William Ames: Dutch Backgrounds of English and American Puritanism*, University of Illinois Press, Urbana.
[32] Trevor-Roper, H.R.: 1956, *The European Witch-Craze of the Sixteenth and Seventeenth Centuries and Other Essays*, Harper and Row, New York.
[33] Williams, G.: 1967, 'Suicide', in Paul Edwards (ed.), *The Encyclopedia of Philosophy*, Macmillan and The Free Press, New York, vol. 8, pp. 43–46.
[34] Williamson, G.: 1969, 'The Libertine Donne', in *Seventeenth Century Contexts* (revised ed.), University of Chicago Press, Chicago, pp. 42–62.
[35] Windstrup, G.: 1980, 'Locke on Suicide', *Political Theory* 8, 169–182.
[36] Wymer, R.: 1985, 'Lodovick Bryskett and Renaissance Attitudes to Suicide', *Notes and Queries* N.S. 32, 480–482.

Oregon State University,
Corvallis, Oregon, U.S.A.

TOM L. BEAUCHAMP

SUICIDE IN THE AGE OF REASON

My objective is to understand how the morality of suicide was framed in the leading writings on the subject in the Age of Reason, with primary but not exclusive reference to the views of Donne, Hume, and Kant. Strikingly diverse attitudes toward suicide developed as Christianity triumphed over paganism, but I shall not explain the different historical forces producing change or survey the variety of viewpoints. Fortunately, the three thinkers on whom I concentrate are representative of important aspects of this development and diversity. Certainly it would be a mistake to group them as libertine or libertarian thinkers. Donne, Hume, and Kant are all lovers of liberty in their own distinct ways, but only Hume holds revolutionary views about suicide that are genuinely libertine.

Although my major concern is with how these thinkers viewed the moral legitimacy of suicide in these thinkers, I begin with some *conceptual* assumptions that were then and still are made about the nature of suicide. Before we can confidently infer that what an author calls "suicide" is what we would ourselves recognize as suicide or that what appears to have been an act of suicide or a practice or policy to handle suicide qualifies for the term "suicide", we need to know *what* we are seeking or think we have found.

I. SOME PERILS OF CONCEPT AND EVIDENCE

The Concept of Suicide

Overdemanding or absurdly narrow criteria of "suicide" would render it impossible to find any suicides or policies regarding suicide at any time. For example, if one were to make it a necessary condition of a suicide that the actor must intend to destroy divine property, it is not clear that any suicide has ever had this intention, even if it were believed that a gift from God (viz., life) was being destroyed. Many conditions similarly would be too strong as necessary conditions of suicide – for example, that no death brought about through the assistance of another person is a suicide or that the means used to bring about death must be an active cause of death rather than a passive

Baruch A. Brody (ed.), Suicide and Euthanasia, pp. 183–219.
© 1989 *Kluwer Academic Publishers.*

permitting of death to occur. Each of these conditions logically excludes acts that would generally be included among the class of suicides.

A death is ordinarily considered a suicide if it is an intentionally caused self-destruction and is not controlled by a coercive or controlling manipulative condition. However, conceptual problems are presented for this elementary definition of suicide by refusals of treatment (with the intent by a person to die from an illness or injury), sacrificial deaths, martyrdom that could have been prevented, actions that risk near certain death or mutilation, self-killings done to escape punishments, and the like. These conceptual problems are compounded by various moral views and psychological theories, and the matter is so complex that an uncontestable definition may be an impossibility. When persons suffer from a terminal illness, mortal injury, or religious fanaticism and intentionally allow their own deaths to occur or give up their lives as a sacrifice for others, many are reluctant to call the act "suicide", sometimes for moral reasons. Similarly, many tend to withhold the label suicide from any act that is not voluntary – for example, acts caused by insanity or by intense pressures from work or family life – or that is not proximately caused – for example, engaging in a lifestyle of drinking or refusing prescribed medication knowing that over a period of years one's death is a near certain outcome.

Others, however, do not share this reluctance about voluntariness. Many psychiatric and legal authorities have held that suicides are almost *always* the result of maladaptive attitudes or maladies needing therapeutic attention. Their conviction is that the suicidal person suffers from some form of disease or irrational drive toward self-caused death. Freudians even argue that suicide is created by a breakdown of ego defenses and a release of destructive forces. But we do not need a psychiatric theory to know that voluntariness is often in doubt in acts of suicide: Many suicides who are not ill may be compromised in voluntariness because they are immature, are unable to process information or deliberate, have fantasies or influential false beliefs, are in a vulnerable position in which they might be manipulated by others, or act under constraints sometimes described as "manipulative situations" or "coercive situations".[1] These conceptual problems were already around in the Age of Reason, when Samson's celebrated act of self-destruction was treated by some defenders of church doctrine as not a suicide and by others as a justified one, thus reflecting a deep conceptual ambivalence in Christian doctrine. John Donne struggled with the status of persons whose actions are contributory causes of their deaths and who express a willingness to die – e.g., "our blessed Savior, who says *de facto*, 'I lay down my life for my sheep'", and others who have "laid down their lives for Christ" or otherwise exposed themselves to certain death for sacrificial or religious reasons ([12], pp. 119–121, 170–176). Donne considered some instances of martyrdom to qualify as voluntary relinquishings of life, and he even cited some partially compelled deaths as suicides – for example, Germanus, who, when thrown to wild animals in the Roman arena, "drew the beast to him, and enforced it to tear his body" (*ibid.*, p. 75).

Others resist linking noble death and suicide. Christian writers have long found difficulty in classifying as suicide martyrs and ascetics who inflict intense privations on themselves that cause death, and these writers have struggled with whether those who kill themselves to protect chastity or honor could qualify (*cf.* [15], pp. 118–133). John Sym, in a book written after (but published before) Donne's, argued that Samson and other martyrs should no more be considered suicides than should a military commander who leads a life-endangering charge into battle [41].[2] Similarly, Kant maintained that intentionally allowing oneself to be killed by one's enemies as a sacrifice for others and in order to do one's duty is no suicide ([24], p. 150). Although only a few of the writers we shall examine provide even an elementary conceptual discussion of "suicide", "self-murder", "self-homicide", and the like, for several of them "suicide" or the nearest surrogate term contains an irreducible *moral* component: An act qualifies as suicide only if it is an instance of unjustified killing where the agent deliberately aims at self-destruction. So value-laden was the concept for some in the early Christian period that not even *killing or allowing the death* of the person was required; some Canons held that self-inflicted castration was sufficient to constitute a self-homicide ([15], p. 126). Donne and Hume make an effort to avoid moral judgments in expressing the meaning of "suicide". Suicide for them is an intentional killing of the self that may or may not be morally justified.

Because understandings of the nature of suicide differ radically, it might be argued that defenders of the morality of suicide and those in utter opposition to suicide have *only* a conceptual dispute, not a moral one, because cases of justified self-killing for the one set of writers are extensionally equivalent with cases of justified self-killing for the other set of writers. A Thomist might argue, for example, that many and perhaps all of Hume's most convincing cases of justified suicide involve persons who lay down their lives for the sake of others, in which case Aquinas might not regard the act as suicide and would therefore not need to condemn it: If a "suicide's" motive is other-regarding rather than self-regarding, and so does not aim at self-destruction, the act might be morally acceptable, and therefore as a matter of *logic* not a case of suicide. If this is a correct interpretation of some who condemn suicide, then they arguably are not refuted by the more libertine opposition that sprang up against them, in the presence of figures like Hume and his French friends, because in the end there is no moral disagreement, only conceptual confusion.

This would be a happy resolution of the many moral problems about suicide that have plagued us for centuries, but it is naively optimistic. Many *moral* problems about suicide, then and now, cannot be waived away by conceptual magic; and the underlying *conceptual* problem is far deeper than the proposed reconciliation suggests, because we have no clear, indisputable meaning of "suicide" – nor was one available in the early modern period that we shall examine and during which the word "suicide" was born. Then, as today, perhaps the major conceptual problem about "suicide" is that the more we have patent cases of actions that *intentionally and actively* cause one's own death, the more we are likely to classify these acts as suicides; but the more the context is one of merely *allowing one's own death when a fatal condition is present or intending something other than to die,* the less inclined we are to classify the act a suicide.

For example, if a seriously but not mortally wounded soldier takes a weapon in hand and intentionally brings about his death (let us presume without an intention of sacrifice or moral assistance to others), it is a clear instance of suicide. But if a seriously injured soldier who is suffering terribly from wounds refuses medical attention, we are not likely to regard his subsequent death as a suicide. The passive nature of the death in the second case makes us reluctant to affix the label "suicide", whereas the active causation of death in the first case makes it easy. Yet suicide does not always involve active steps. If the soldier suffered only a minor wound that could easily be repaired, yet refused assistance with the explicit intention to end it all, the intention to die may overshadow the passive route chosen to death in our conceptualization of the act. Other cases will be on the borderline where we are not sure what to say, because of the nature of the means chosen and the character of the intention involved.

In order to avoid as many of these problems as possible, and especially to avoid prejudicing the moral analysis of suicide by excluding altruistic or divinely inspired actions, I shall adopt what I believe to be an untainted and simple definition of suicide throughout this essay: Suicide occurs if and only if *one intentionally terminates one's own life* – no matter what the conditions under which the act is performed (e.g., one can be coerced) or the precise nature of the intention to terminate (e.g., one need not wish to die) or the causal route selected to bring about death. If a soldier's refusal of treatment is made with the intention to terminate life and does so successfully, the act is a suicide; but if the soldier is merely giving up his allocation of a medication in order that a companion might live and has no intention to end his life by the action, then his subsequent death is not a suicide. There are, of course, problems about the nature of the distinction between the active and the passive and about what qualifies as an intention to terminate life, but these problems plague virtually all definitions of suicide and cannot be further examined here (*cf.* [6], esp. pp. 78–89, and [4], pp. 33–68).

I will rarely mention or make specific appeal to this streamlined, broad definition, but at least it provides advance notice of what I am attending to when I judge whether a particular act is a suicide. Unfortunately, on some occasions my definition will not recognize as a suicide what a given author under discussion unhesitatingly considers a suicide – for example, cases in which self-caused deaths occur where an actor hoped to survive a tragic situation but failed.

Historical Evidence

Sometimes problems of assessing suicide turn not on the adequacy of the conceptual criteria being used as they do on the *evidence* available to determine whether these criteria were satisfied. We shall see, for example, that John Donne resists evaluating the moral quality of the Biblical suicides because of the difficulty in *determining* the motives of the actors. During the Age of Reason, meaningful and reliable evidence from old court records, medical journals, and the like is virtually nonexistent. The bulk of evidence of what we know about the occurrence of suicide is from poorly documented anecdotes, sermons, Scriptural commentaries, moral tracts and poems, and occasional eye-witness reports motivated by reasons that often irreparably taint the reports – e.g., public accounts of a suicide note accompanied by accusations of the person's premeditated murder against himself or herself ([40], p. 35).

A related question regards what can be reasonably inferred from published lectures, general pamphlets, and treatises, which were usually written by prominent citizens of the religious or academic communities, but which may not have been representative of community beliefs and practices, especially in non-Christian communities. In the absence of more direct data about actual suicide practices and sanctions, these documents are unreliable sources of information about community sentiment. It is not always clear whether the statements made in these documents were primarily exhortatory, descriptive, or self-protective: some writings describe, for educational or deterrent purposes, conduct that is exemplary or evil rather than representative or borderline. Moreover, despite statistics kept in some periods, most of the information retained about suicide is suspect. For example, it is known that both family members and jurors attempted to protect families from judgments regarding suicide, because suicide was, according to the law, a *felo de se,* and a suicide's chattels had to be forfeit to the king. Any suicide that could be covered up therefore likely was ([16], p. 106). We also still have but poor information about the extent to which group or mass suicide occurred in the Puritan community and the Jewish community, as well as about the justifications that might have been invoked for them. The paucity of available evidence of suicide therefore invites interpretative and historical dispute.

I am not arguing the skeptical thesis that we can know nothing reliable about suicide in the Age of Reason. I am only warning that historical evidence about cultural beliefs and practices is tenuous. Fortunately, this problem does not present insurmountable difficulties, because we are concentrating on the intellectual history of major writings.

II. THE CULTURAL BACKGROUND OF PHILOSOPHICAL AND RELIGIOUS BELIEF

By the time Western culture emerged as Christian culture at the end of the Middle Ages, religious attitudes toward suicide were, on the available evidence, well entrenched through the intellectual writings of the Church fathers and the social control exerted by the Church. Many Christians had in previous centuries committed suicide to avoid the sins of this world, and the response that ultimately formed in the Church was that the sin of suicide is graver than any sin a suicide could hope to avoid or repent by the act. Three Church councils in the sixth century solidified a rapidly growing opposition in the Church to suicide, and the opposition thereby became culturally effective within the Christian community. However, outside that community – among heretics, Jews suffering from pogroms, and excommunicates – the same effectiveness has not been documented.

The Religious Defense

Although we are studying philosophical rather than cultural history, we should not lose sight of the religio-cultural shield that kept libertarian arguments from penetrating religious defenses well into the nineteenth century. The pivotal points in this shield are two: (1) theological views about scripture and divine command, and (2) related views about the value of life and about moral courage and failure in confronting life's ills.

(1) According to the prevailing theology, suicide and other acts are morally wrong because they violate a direct command of God against the taking of human life. The claim is defended by scriptural or theological appeals – rather than by appeals to philosophy, culture, or the authority of tradition. Metaphor is commonly at work in the arguments – e.g., that human persons (or perhaps their souls) receive the gift of life from God, and therefore suicides sin against their creator by the act of destroying their lives. An account of divine command commonly supplies the underlying premise: Because murder is specifically prohibited by God and because suicide is self-murder, suicide is therefore morally wrong unless directly commanded by God.

Many theologically based arguments against suicide turned at the time on an account of some special design of providence, e.g., that suicide violates an obligation to God by interfering with the divinely ordained order of the universe. It is not always clear what God's "order" is, but the notion of God's appointing a special purpose for each human life is prominent. These arguments rest on theological rather than philosophical grounds, and therefore theological principles function to determine the morality of the act of suicide. If these justifying reasons are held to be *valid independently* of philosophy, then philosophical objections can make no dent. The prevailing Christian view on suicide seems to have been that no philosophical counterargument ould count against the prevailing theological premises.

(2) The reigning theology led to a view about the absolute value of life: In consonance with Christ's teachings, human life has value in itself, and suicide is wrong because it destroys something of inherent value – that is, a life that is valuable not for any other value it brings into existence but simply because of its own inherent value. This principle was commonly construed so that it is, under some conditions, permissible to allow someone to die or to permit and in some cases even encourage one's own death, instead of attempting to preserve or save life.

The Legacy of the Church's Philosophical Theologians

The most influential philosophical and religious arguments against suicide throughout the period under examination were the Augustinian and Thomistic theories that informed the Church's position. These arguments directly parallel the content of the major essays that we shall examine, in particular Donne's and Hume's. Virtually everything asserted by the Church's philosophical theologians is disputed by later secular thinkers such as Hume, and most of it is challenged even by Donne. At the bottom of this debate, we find conflicting principles from profoundly distant world views and social theories.

St. Augustine's (A.D. 354–430) writings on suicide had already established the basic outlines of the Christian position almost a thousand years before Aquinas wrote: Scripture condemns suicide through the sixth commandment not to kill, and therefore only a countercommand by God, as in the case of Samson's self-caused death, qualifies as a justification of suicide ([2], pp. 17–27, esp. 21–22, 26).[3] Augustine regarded suicide as cowardly in escaping the ills of life and thought the act deprived the suicide of the opportunity for repentance. Although his views remained influential for centuries, St. Thomas' views ultimately prevailed at the pinnacle of orthodoxy, for complicated reasons including Aquinas' own authority, the detail in his arguments, and his use of natural law theory. St. Thomas argued that suicide is an offense against self, an offense against society, and a violation of God's sovereignty and ownership of human life. Although St. Thomas does not rely on scriptural authority as heavily as others, a vital premise in the argument is that suicide runs counter to the creator's general interest in human life and its flourishing. Here are condensed versions of the wording in Aquinas' arguments, as found in the *Summa Theologica* 2–2, 64, 5:

It is altogether unlawful to kill oneself, for three reasons:

1. because everything naturally loves itself, the result being that everything naturally keeps itself in being... Wherefore suicide is contrary to the inclination of nature, and the charity whereby every man should love himself. Hence suicide is... contrary to the natural law and to charity.
2. because... every man is part of the community, and so, as such, he belongs to the community. Hence by killing himself he injures the community...
3. because life is God's gift to man, and is subject to His power. ... For it belongs to God alone to pronounce sentence of death and life...

The first Thomistic argument (1) rests on premises of natural law, which may be reconstructed in the following form:[4]

1. It is natural law that everything loves and seeks to perpetuate itself.
2. Suicide is an act contrary to self-love and self-perpetuation.
3. (Therefore) Suicide is contrary to natural law.
4. Anything contrary to natural law is morally wrong.
5. (Therefore) Suicide is morally wrong.

St. Thomas is here building a distinctive natural law theory on the foundations of a Stoic theory (Aquinas may have thought it was Aristotle's, to whom it is still today wrongly attributed). The theory is that the good of any creature in nature consists in the actualization of its distinctive natural potentialities. The good of a snail consists in the actualization of its distinctive natural potentialities, and the good of a human person consists in the fulfillment of its (quite different) complement of natural potentialities – i.e., those properties distinctive of human persons. All living organisms flourish if and only if their parts function harmoniously as they should – where "should" is understood in terms of the way nature dictates or "legislates". A person's body, mind, and emotions must be coordinated and be in accordance with natural functions in order to be healthy: One is mentally and physically healthy only if the mind and body function as nature has legislated, and one is morally healthy only if one attempts to fulfill one's own potentialities as well as the potentialities of others.

Certain human inclinations are selected in the Thomistic theory as the basis of natural laws, while inclinations that destroy the possibility for the actualization of basic potentialities are depicted as unnatural deprivations.[5] Suicide is wrong in this theory precisely because it violates a natural inclination to the conservation of existence and well-being (and, as we shall see, because the suicide does not exclusively intend a good outcome). This theory permits the Thomist to acknowledge the psychological fact that powerful inclinations to suicide are "natural", while denouncing them as, in a different respect, unnatural deprivations.

Premise (2) seems to commit Aquinas to agreement with later secular thinkers that some suicides are rational and avoidable. Many philosophers and psychiatrists would deny this rationality thesis, and yet would continue to use an argument similar to (1)–(3). Premises (1)–(2) are acceptable to them if and only if (1) is construed as a law of human nature denying that there can exist a *natural* internal impulse to self-destruction, although there can be forces "external" to the self – such as physical causes or influences creating psychological compulsion – that cause suicide. Unfortunately, these distinctions are obscure in Aquinas and remain underdeveloped throughout the period that we shall be examining.

Also obscure in Thomas' writings is the nature of the essential distinction between laws of nature and natural laws, only the latter of which play a role in his moral argument. Laws of nature presumably are descriptive statements derived from scientific knowledge of regularities in nature, while natural laws are normative statements derived from philosophical knowledge of the essential properties of human nature. In this theory, natural laws do not empirically describe behavior; rather, they delimit the behavior that is morally appropriate for a human being, and so tell us how we ought to behave because of our very nature as humans. What is proper for a human differs from what is to be expected from other creatures insofar as their "natures" differ; and their natures differ because they possess different essences with different potentialities. For example, while humans share with animals a natural tendency toward sexual reproduction, only through human reason can there be a tendency toward universal goods, such as concern to promote the interests of others.

Aquinas' second argument [2] is elegantly simple, by comparison with the first. This social-obligations argument asserts that every individual belongs to family and neighbors and has obligations that are violated by an act of suicide. While Aquinas does not believe that society is sovereign over the individual, he does believe that individual decisions cannot be properly made in even so personal a matter as suicide without reference to the interests of other persons in the State. He regards an act of suicide as undermining social authority and human relations and therefore as harming all those affected by the action.

Aquinas' third argument [3] exerted massive influence among Christian believers. According to this argument, human life is a gift from God, and all individuals belong to their Creator just as a piece of property belongs to its owner. To commit suicide is tantamount to the sin of theft, because it deprives God of that which is rightfully His. This argument presupposes that it is the will of God that we remain alive, despite cases such as Samson's. Aquinas also apparently thinks that the suicide necessarily intends evil effects except when specifically commanded by God to perform the act.[6]

Aquinas' theologically-based set of objections to suicide reigned supreme despite the Renaissance's wave of respect for autonomy that weakened Church authority over the individual. Defenses of suicide – often within the framework of Christian belief – began to emerge through suggestive statements made by Erasmus in *Praise of Folly* (1509) and Sir Thomas More in *Utopia* (1516), but these works were neither comprehensive nor systematic treatments of suicide; and they made no dent in cultural or religious beliefs. The first book to achieve such status was John Donne's *Biathanatos*.

III. DONNE'S BIATHANATOS

When *Hamlet* was composed in 1600, suicide seems to have been as condemned and feared in the Christian world as any natural disaster, disease, or anti-social act ([40], p. 2, n. 1). A typical example of the prevailing attitudes of the time is found in the 1594 lectures of John King, later bishop of London. King had taught that Scripture delivers "a commaundment in express tearmes" against suicide and had argued from the premises proposed by St. Thomas: suicide is wrong by nature and is a violation of obligation to both God and society. Similarly, George Abbot, later Archbishop of Canterbury, lectured in 1600 that suicide was categorically forbidden by the sixth commandment and that we are absolutely obligated to follow these commandments (*ibid.*, pp. 3–5).

These were the cultural waters into which Donne sailed when, at approximately age 36, he composed *Biathanatos* (1607–08). Known only to a small cadre of friends, the book was posthumously published by his son more than a decade after his death.[7] It is a work so difficult to interpret, that some have hypothesized that Donne did not intend to defend suicide ([20], [10]). Even his motivation to authorship is uncertain: he seems to have suffered from a melancholy that motivated his interest in suicide, but his apparent purpose in writing the work was intellectual and sprang from his interests in theology and law (*cf.* [13], pp. ix–xiii). In both his preface and conclusion he states his intention to discuss self-homicide in order to combat an ignorance of mind that unreflectively condemns it. He sought to elevate an emotionally tainted subject to an intellectual plane of discussion, assessing the motives of suicide, situating those motives in the proper place in the mind, and assessing the acceptability or unacceptability of suicidal actions and intentions.

This objective may seem paradoxical in light of the meaning of the title, which is "violent death" or "dying a violent death". Our word "suicide" is a later seventeenth century creation that despite its Latin roots has no clear equivalent in either Greek or Latin. Donne chose "biathanatos" because of a prior Greek usage of this term that was applied to the kinds of cases that we now call "suicide", but nonetheless seems to refer to intentional self-inflicted death brought about by a person's action. Donne was an author who dwelled on paradox, and whatever the reason for the title, it was the first book in English to defend what we today call "suicide" – or what he typically terms "self-homicide". The subtitle placed on the work gives a more synoptic picture of its objective and argument: "A Declaration of that Paradox or Thesis, that Self-Homicide is not so naturally Sin that it may never be otherwise; wherein the Nature and the Extent of all those Laws which seem to be violated by this Act are diligently surveyed" ([12], p. 1).

Donne had been advised by friends not to publish the work and was both ambivalent about eventual publication and about whether the manuscript should be destroyed. It is not known, however, whether he esteemed his friends' advice or why he was ambivalent about publication. Fear of ecclesiastical sanction is an intuitively appealing explanation, but this account is unsupported by the available facts. Quite the reverse: It is possible that Donne had ecclesiastical and consequential worries of his own that the book would do more harm and cause more sin than it would alleviate (*cf.* [12], pp. x–xvi, lxxxiv).

Although he says that "my end is to remove scandal" (*ibid.*, p. 44), his is neither a sweeping nor a secular defense of suicide. The argument is only that under broader conditions than had previously been appreciated in the Christian community, suicide may be justified and may escape the category of sin. In his sermons Donne generally condemned suicide (*cf.* [13], p. xxxii, n. 20), and in *Biathanatos* he notes that suicide is wrong in many circumstances – e.g., where self-interest predominates, an intent to prevent other persons from sinning may not be sufficient reason for Donne ([12], p. 102). Yet he clearly disagrees with Augustine and like-minded theologians about the *range* of intentions and circumstances that justify suicide. A major premise in Donne's position is that possession of the intent to act for the glorification of God is sufficient; it is not essential that there be an actual summons by God.

Donne's book flowed by design directly from the structural organization of St. Thomas's tripartite definition of sin: "Through [Aquinas'] definition, therefore, we will trace this act of self-homicide, and see whether it offend any of those three sorts of law", viz. "these three laws of nature, reason, and God" (*ibid.*, p. 51). The three major parts of Donne's essay are divided, for this reason, into an examination of laws of nature (natural law), laws of reason (positive law), and laws of God (divine law). To this extent his response is *directly to Aquinas*, who had fashioned his own discussion of suicide similarly in accordance with the tripartite definition of sin. The logic of Donne's organization and argument seems to be to show that the traditional division for defining sin will not prove adequate, as Aquinas thought, to show that suicide is always sinful or illicit.[8]

The Attack on Natural Law Condemnations (Part I)

In Part I Donne concentrates on St. Thomas's first argument, from natural law. Donne first shows that the notion of a natural law is systematically unclear, exhibiting several distinct meanings; but none of the variant meanings precludes the legitimacy of *some* acts of suicide ([12], pp. 52–60). Apparently in direct opposition to Aquinas, Donne argues that while some persons naturally strive to keep themselves alive, there is also a "natural desire of dying" bred into us by nature. This strikingly original thesis – found virtually nowhere else in the literature of suicide before or since – is defended by an examination of a wide variety of actual suicides performed from an equally wide variety of motives, both within and without Christian culture (*ibid.*, pp. 63–68).

One reason Donne found the natural-desire thesis congenial was the Christian belief in an afterlife. Donne did not view self-killing as total self-annihilation: Every believing Christian would desire to pass from this life to the hereafter, and therefore it is hardly unnatural for a true believer in a community of true believers who reinforce belief in the afterlife to desire death as a means to this blessed outcome: "Heaven, which we gain so, is certainly good; life, but probably and possibly... If that which I affect by death be truly a greater good, wherein is the other, stricter law of nature, which is rectified reason, violated?" (*ibid.*, pp. 63–64). The reason, for example, why church fathers like Tertullian and Clement of Alexandria could invite and even urge martyrdom, Donne reasons, was the prior presence of a natural inclination on the basis of which persuasion to martyrdom was possible (*ibid.*, pp. 70–71).

Donne therefore ventured the conclusion in Part I of his book that the naturalness of suicide makes it in accord with, rather than opposed to natural law, and that suicide is therefore neither sinful nor blameworthy unless our natural desire for death is perverted or corrupted into a desire to die for wrongful reasons. In Donne's judgment, the Thomistic conception of natural law was too narrowly conceived and for this reason failed to understand how there could be morally defensible suicides (*ibid.*, p. 83). Donne's view is that many desires, such as sex and hunger, are natural and become wrongful only if corrupted. Although the claim that suicide is natural never inspired subsequent writers, the challenge that suicide is not unnatural was profoundly important for later theories, because it stood as a fundamental challenge to Thomistic natural law prohibitions on suicide. This may not have been Donne's intent, but it was the eventual outcome.

The Critique of Community-Based Criticisms (Part II)

Although Part II of *Biathanatos* is formally on the "law of reason" and laws in the state, the arguments parallel and answer St. Thomas's second argument, that a suicide injures the community. Donne acknowledges that there are civil and religious laws against suicide that attempt to deter it and that a widespread practice of suicide could reduce the population and even endanger the existence of the state, but he argues that these laws may only work to coerce a person to long-suffering who otherwise might have justifiably acted on the natural desire for death. All of these general rules, he argues, require interpretation by human reason, and often the civil and criminal laws rest on no more than what is customary at the time. Donne is not opposed to having laws if they are essential for the state's protection. But he notes that even the presence of good reasons for anti-suicide provisions in the Civil and Canon laws in no way indicates the *moral* wrongness of the act prohibited (*ibid.*, pp. 96–100).

Donne is open to agreement with Augustine and Aquinas that suicide often is not morally permissible and that anti-suicide statutes can be justified, while at the same time maintaining that there is nothing intrinsically wrong with suicide. That is, suicide is not *by its nature* sinful or unforgiveable. Donne then argues that suicide can be justified if it is done with the intent to glorify God and without an intent at self-promotion. He tries to show that Augustine thought otherwise only because he consistently failed either to consider the right kinds of examples or to draw proper conclusions from the examples he uses such as those of Samson and the virgin martyrs. Donne's position is that judgments of the morality of action ought to be made not on the basis of *consequences* to the community but rather by assessing the *intention* of the actor. In these passages Donne defends conscience, which for him denotes the faculty of reason (and thereby the voice of God in the person) through which we test whether our motives are proper. In the case of suicide we judge whether it is done for the glory of God rather than for self-benefit (*ibid.*, pp. 102–103, 136–137).

A Reinterpretation of Scripture and Christian Life (Part III)

In Part III Donne addresses a range of claims that arise in the third Thomistic argument, which is based on the premise that human life is a gift from God to be disposed of exclusively by divine initiative. However, Donne concentrates less on the Thomistic claim than on an argument to show that no law of God need be violated by suicide and that suicide is not condemned or banned by scripture. Because scripture is the medium of God's message, this approach is essential for the understanding of what God has done in the way of establishing divine guidelines. Donne argues that whereas commentators disagree over the meaning and sanctions of scriptural passages in general, in the specific instance of suicide there is no overriding scriptural prohibition, nor any single case where an act of suicide is condemned. He argues that even the most frequently cited passage – the commandment not to kill – is generally recognized to have exceptions, of which suicide is never said in scripture not to be one (*ibid.*, pp. 147, 154).

Donne is challenging one of the sleights of hand of the period: the ensconced view that Scripture itself condemns suicide. Donne noticed that the Old and New Testament do not forbid suicide and that there is at least as much acceptance of suicide as rejection in the accounts given. Of the suicides found in the Old Testament – Samson, Saul, and Achitophel – none is condemned (*ibid.*, pp. 180–187). When Judas commits suicide in the New Testament, the textual treatment is as much that of rightful repentance as criminal behavior (*ibid.*, 186–187). Donne, then, was challenging the orthodox belief in scriptural prohibition using "empirical" evidence from scripture and history: Not until after Augustine's time had the Church – without clear scriptural warrant – ruled that suicide was illicit and turned to scripture for authority to back the claim. Donne vigorously protests Augustine's strategy of assuming axiomatically that Biblical suicides by praiseworthy characters must have been instances in which God commanded the self-inflicted deaths. This strategy is viewed as but a convenient way of begging the question. Donne's approach is not to pass judgment on these suicides at all, on grounds of his earlier argued position that the person's intention alone counts. Because intentions are hidden, we are in a difficult position when it comes to moral evaluation. Even Judas, who has perhaps the clearest motive, is not evaluated by Donne, and for principled reasons.

In Part III Donne also argues that it is no part of God's law, dominion, or design that suicide is necessarily a violation of God's gift of life. Only if there is disobedience in the use of the gift does an act become sinful and a moral wrong worthy of condemnation. Thus, human reason and conscience might be in accordance with God's law rather than in opposition.

Post-Biathanatos: Reactions and Retreat

As a counterfactual we might speculate that Donne's work would have passed unnoticed had it not been the case that he eventually emerged as dean of St. Paul's in London. But he did come to occupy this position, and his challenge to ecclesiastical authority on the issue of suicide did receive considerable attention. This challenge did not make significant headway against the religious and intellectual currents of the time, and found not even a minor figure to rise in its defense. At the time Donne began to revive ancient views about the acceptability of suicide, the arguments were received by the culture more with revulsion than reflection.

To the extent there was a scholarly community at the time, it did not rush to Donne's defense either to protect academic freedom or to legitimate suicide. Dozens of tracts under the influence of Anglican and Puritan beliefs dominated the landscape of publications, and many were implicitly or explicitly critical of Donne. Not two years had passed after *Biathanatos* was published when Henry Hammond began to hammer away at Donne:

> For power over a mans owne *life* no man can be believed to be borne with it, for if he were, he might then as lawfully *kill himselfe* (and if he might do it *lawfully*, there are many cases which might make it *prudent* for him to do it at some times) as put out an aking tooth, or dispose of his liberty or estate out of his owne possession; a thing which no Christian hath ever thought lawfull, but made this *felinia de se*, this self-murther, or felony against ones selfe, a crime, as contrary to the sixth Commandement, as the *killing of any other man*... Before... the posthumous *Biathan*: was set out, a man might in this Nation have adeventur'd to... not have feared a challenge in *print* for his injury... (quoted in [40], p. 57).

The following views expressed by Jeremy Taylor typified the Anglican viewpoint, and it is not difficult to see how little Donne's views had influenced his peers to abandon Aquinas:

> It is impiety and rebellion against God; it is desertion of our military station, and a violation of the proprieties and peculiar rights of God... It is against the law, and the voice, and the very prime inclination of Nature. Every thing will preserve itself ([42], vol. II, pp. 71 ff.).

In writing after writing in the post-Donne era suicide is condemned as murder, contrary to religion, forbidden by God, criminal behavior, against nature, against human nature, sinful, against virtue and morality, and against Scripture ([40], pp. 41 ff.). Donne's work had done little to challenge the moral certainty and rectitude that underlay these attitudes. This is not to deny that secular thought was on the rise in a world finding increasing difficulty in sustaining Augustine's distinction between a city of God and a city of man, whereby the latter was to be ruled by the laws of the former. But it is still a long stride from the breakdown of this belief to a breakdown in religious authority over such matters as suicide, salvation, and sin. In the late seventeenth century Donne's voice may have been the most prominent and influential yet to arise in defense of suicide, but it was only a weak voice both in its message and in its cultural impact – at least weak by comparison to a much smaller essay whose impact I shall now consider.

IV. HUME'S ESSAY

In his essay "On Suicide",[9] David Hume presented the best developed and most perceptive set of arguments for the moral permissibility of suicide in early modern history, although many of the views he advanced had already been adumbrated in the Stoics, Donne, and Montesquieu. Like Donne, Hume expressed a desire that his essay be suppressed after an early distribution of the essay had been privately arranged. His words and actions indicate that he had serious reasons for suppression, but, as with Donne, Hume's reasons for suppression have never been satisfactorily explained. Some scholars have argued that religious authorities threatened and intimidated Hume and his publisher, and that for political and prudential reasons they capitulated when the pressure became intense and when friends like Adam Smith advised against publication ([31], pp. 37–44, 52f.). Although the fact of attempted intimidation is beyond reasonable historical doubt, that the intimidation was successful and constituted Hume's reason is based on hearsay accounts written by enemies of Hume, such as William Warburton. Hume himself says only that he suppressed the essay "from my abundant prudence" and under advice of a friend ([21], p. 253). An alternative and plausible hypothesis is that he had reservations about the originality of the essay or about the quality of the arguments.

However, there certainly were good reasons to be cautious about what one published on the subject at the time. Ecclesiastical penalty, customs of dishonoring the corpse, and legal sanctions were all still in force – practices recorded at the time in Blackstone's *Commentaries* (vol. IV, p. 190).

A typical example of an influential contemporary theological viewpoint is found in the following arguments by the reverend and learned Isaac Watts, D.D.:

In the rubric of the church of England, before the burial service, self-murderers are ranked with excommunicated persons; the church has no hope of them as true christians; and as the church denies them christian burial, so the civil government did heretofore appoint that they should be put into the earth with utmost contempt; and this was generally done in some public cross-way, that the shame and infamy might be made known to every passenger; and that this infamy might be lasting, they were ordained to have a stake driven through their dead bodies, which was not to be removed. It is pity this practice has been omitted of late years by the too favourable sentence of their neighbours on the jury, who generally pronounce them distracted; and thus they are excused from this public mark of abhorrence. Perhaps it were much better if this practice were revived... to testify a just abhorrence of the fact, and to deter survivors from the like practice ([44], p. 476).

Hume did not deal directly with such theological and cultural attitudes in the essay. Rather, he chose the medium of philosophical discourse about classical arguments.[10]

In the way of critique, Hume offered a battery of arguments against theological views such as those expressed by St. Thomas. Unlike Donne, Hume does not explicitly name Aquinas as the inspiration of his essay; and there were plenty of contemporary targets at which Hume might have been aiming. For example, just before Hume's birth the well-known philosopher and divine Samuel Clarke had argued against suicide (as well as death by debauchery and self-abuse), using arguments from eternity and immutability of the laws of reason and moral obligation (9). (Clarke's ethics was one of Hume's favorite targets in his general moral theory.) Although Hume never mentions Aquinas, the arguments he attacks are the Thomistic arguments or close approximations of them. Hume attacks the three arguments advanced by Aquinas and only those three arguments, albeit in heavily modified language. No other prominent historical source known to me had used these three and only these three arguments against suicide. Somewhat similar arguments had been used by Locke and Warburton – as well as by numerous other philosophical and theological figures. But the Thomistic triad of arguments seems to be Hume's aim and I shall therefore speak of the arguments he attacks as Thomistic arguments.

As a constructive argument in defense of suicide, Hume appealed to the autonomy rights of individuals and to the generally good consequences of some instances of self-caused death. Although his essay is an entirely secular justification of certain types of suicides, Hume did not draw the radical conclusion that all suicides can be justified merely because they involve autonomous choice. His strategy was to show that the more one's life is plagued by suffering or some related reason for ending it all, and the less one has outstanding obligations to others, the more justifiable is the act of suicide. In the end, he advanced the largely utilitarian thesis that if the value of relieving one's misery or dire circumstances by taking one's own life is greater than the value to the community of one's continued existence, then suicide is justified. However, some of his arguments suggest that one has the right to commit suicide under some conditions without considering utility to the community.

The main business of the essay is stated in straightforwardly libertine terms: "Let us here endeavor to restore men to their native liberty, by examining all the common arguments against suicide, and showing that action may be free from every imputation of guilt or blame..." The remainder of Hume's essay is an intricate polemic against the proposition that, "If suicide be criminal [immoral], it must be a transgression of our duty either to God, our neighbor, or ourselves ([23], p. 152). Throughout the essay, Hume identifies most closely with classical writings, especially those of the Greek and Roman Stoics. For example, he accepts the view that life is not worth living if ill health seriously and irreversibly undermines one's potential for enjoyment. Hume was a classicist and a secular philosopher with opinions that were more congenial to ancient culture and intellectual life than to those thriving among his Christian contemporaries and predecessors.

The Liberty-Based Arguments

Hume's Positive Arguments. Hume's major constructive argument is based almost exclusively on moral considerations of utility and respect for autonomy: He first tries to show that in some cases resignation of one's life from the community "must not only be innocent, but laudable". His strategy is to analyze hypothetical cases that not only present moral reasons in favor of suicide but that also stand as counterexamples to St. Thomas's claim (in his second argument) that "by killing himself [a person] injures the community".

Hume begins with an analogy to the rights of retired persons: Suppose a man retires from his work and from all social intercourse. He does not thereby harm society; he merely ceases to provide the benefits he formerly did by his productivity and amiability. Hume here relies on a general claim about the reciprocity of obligations:

> All our obligations to do good to society seem to imply something reciprocal. I receive the benefits of society, and therefore ought to promote its interests; but when I withdraw myself altogether from society, can I be bound any longer? But [even] allowing that our obligations to do good were perpetual, they have certainly some bounds; I am not obliged to do a small good to society at the expense of a great harm to myself: when then should I prolong a miserable existence, because of some frivolous advantage which the public may perhaps receive from me? (*ibid.*, p. 158).

Hume next considers a string of hypothetical and actual cases of suicide. Each case in the string contains a new element not contained in previous cases that increases the personal or social value of death for the suicide. In his first hypothetical case, Hume envisages a sick, but still marginally productive person. If his social contribution is small in proportion to the largeness of his misery, then Hume thinks there is no social obligation to continue in existence. The presiding principle is again utilitarian: one's misery overrides value to the community.

The second hypothetical case in the potential suicide's existence is so bleak that he is not only miserable and relatively isolated from social contact but entirely a burden to society. In the third, a political patriot spying in the public interest is seized by enemies and threatened with the rack, and is aware that he is too weak to avoid divulging all he knows.[11] In both cases, Hume stipulates that these unfortunates shall remain miserable for the remainder of their days. He then proclaims acts of suicide under such conditions praiseworthy because they satisfy both the individual's primary needs and promote the larger public interest. These examples are intended to show how persons might be so situated that everyone benefits from their suicide. Whether the State is advantaged or disadvantaged by citizen involvement is relative to the citizen's situation, and Hume thinks that there are conditions under which suicide not only promotes the interest of the individual but honors and shows respect for the person's family. He even maintains, dubiously, that "most people who live under any temptation to abandon existence" act from such utilitarian motives (*ibid.*, p. 159).

The Anti-Theological Arguments

In his opening lines Hume warns that religious foundations of moral beliefs about suicide are inadequate for establishing a reasoned position. The job of philosophy is to challenge the presuppositions of religion, superstition, and human psychology. Hume suggests that the fear of death and superstition partially *explain* belief in the immorality of suicide, but these phenomena will not suffice as a *justification* of belief. With this warning of the priority of philosophy over theology, Hume considers the traditional Thomistic arguments. Whether he intentionally directed his views against St. Thomas is not known. These arguments were so prevalent in law and theology, ([36], pp. 234–235) that he need not have had any one person in mind.

Hume's strategy is not to challenge belief in the existence of God, but to isolate a general theological proposition for critical investigation, viz., *the act of suicide violates an obligation to God and provokes divine indignation because it encroaches on God's established order for the universe.*[12] This proposition captures Aquinas' third argument, albeit in reconstructed language. Hume gradually substitutes alternative of "God's established order". He then argues that on the particular interpretation in question, the theology is either deficient or is compatible with the moral acceptability of suicide. I shall refer to the three main theological interpretations that he rejects as (1) The Divine Ownership Interpretation; (2) The Natural Law Interpretation; and (3) The Divine Appointment Interpretation.[13]

(1) The Divine Ownership Interpretation. Hume opens by considering the possibility that encroaching on God's established order is wrong because "the Almighty has reserved to himself ... the disposal of the lives of men, and has not submitted that event ... to the general laws by which the universe is governed" ([23], p. 154). This interpretation parallels Aquinas' claim that "it belongs to God alone to provide sentence of death and life," while death "is subject not to man's free will but to the power of God." Plausibly Aquinas' first argument presumes that human life is God's property; He created it and retains rights of control over it. Hume denounces this right-of-disposal contention as "plainly false", because human lives, like all matter in the universe, are subject to general causal laws. Because persons die of natural causes – as in the cases of being poisoned or swept away by a flood – it is gratuitous to maintain that there is a divine cause. That is, because such events are explicable by general causal laws, and without reference to divine volition, appeal to divine intervention is gratuitous. Hume also argues against the divine ownership thesis by maintaining that, "If my life be not my own, it were criminal for me to put it in danger, as well as to dispose it..." (*ibid.*, p. 156).

(2) The Natural Law Interpretation. In a second search for the meaning of "encroaching on God's established order", Hume focuses on the claim that it is wrong to disturb the operation of any general causal law, because the

totality of causal laws constitutes the divine order. Hume construes this "natural-law" thesis to mean that human beings morally should be absolutely passive in the face of natural occurrences, because otherwise they would disturb the operations of nature. Hume rightly ridicules this theology (if anyone ever held it) as absurd: Unless we resisted some of nature's forces by counteractions, we "could not subsist for a moment," because the weather or some other "natural" event would destroy us. But if it is morally permissible to disturb some operations of nature, then it is morally permissible to avert life itself by diverting blood from its natural course in human vessels. This action relevantly resembles turning one's head aside to avoid a falling stone, because both divert the course of nature. "It would be no crime in me to divert the Nile or Danube from its course," says Hume. "Where then is the crime of turning a few ounces of blood from their natural channel?" (*ibid.*, p. 155). This argument may seem shabby and misdirected as an argument against natural law philosophers, who draw a distinction between laws of nature and natural laws. However, this distinction is too obscure to be convincing to Hume, who may have believed that a major defect in the argument is that it rests on such an arbitrary distinction. Unfortunately, he does not *argue* for his claim, perhaps because he could see no reasons either for or against such a claim.

(3) The Divine Appointment Interpretation. Hume's final accounting of "encroaching on God's established order" rests on the theological view that nothing in the universe happens without providential "consent and cooperation" (*ibid.*, p. 157). He outlines this interpretation as follows: "You are placed by Providence, like a sentinel, in a particular station; and when you desert it without being recalled, you are equally guilty of rebellion against your Almighty Sovereign, and have incurred his displeasure" (*ibid.*). With this interpretation – the only sentence italicized in full sentences in the entire essay – Hume strikes at the heart of the theological matter. Hume's "rebellion" is Aquinas' "sin". The argument had also appeared in Locke, and would later appear – as we shall see – in strikingly similar form in Kant. But its popularity fails to impress Hume. He pronounces the argument absurd as an objection to suicide, because if this measure of divine causal control be exerted, then nothing in human life happens without divine consent and "neither does my death, however voluntary, happen without its consent." He concludes with an appeal to the argument (which is broader than Donne's) that whenever I no longer have a wish to live because of pain and exhaustion, and I want to die, it seems fair to "conclude that I am recalled from my station in the clearest and most express terms" (*ibid.*).

Hume also offers – almost offhand – an interesting anti-theological argument based on the importance of motives. Quoting Seneca, he observes that it is consistent that a man should take his life in virtue of his misery, while at the same time expressing sincere gratitude to God "both for the good which I have already enjoyed, and for the power which I am endowed of escaping the ills that threaten me" (*ibid.*, p. 156). This one-sentence suggestion, so reminiscent of Donne's arguments, constitutes an important objection to Aquinas by calling into question the supposition that God has willed that we should never take our lives in virtue of our misery, irrespective of attitude and intention. Hume's argument, again more sweeping than Donne's, is that removal of misery is a truly good effect and the intention to alleviate misery cannot by itself be a condemnable motive, even if suicide is the unfortunate means to this praiseworthy end. Additionally, it cannot be regarded as evil or sinful in intent if accompanied by a sincere expression of gratitude to God.

Post the Posthumous Essay: Reaction to Hume's Arguments

The reaction to Hume was slower than might have been expected, and it was not until the first authorized English edition (of 1783, after an unauthorized edition in 1777) that there was a visible reaction. At this point the responses were swift and not unlike the attention that Donne's essay had received 130 years prior. In 1784 the Right Reverend George Horne, Lord Bishop of Norwich, wrote a sustained response containing a sentence-by-sentence refutation.[14] He ridiculed Hume's assumptions and arguments as mischievous, as constituting an end to human morality, as "a bad example of impatience and despair", and as "against the voice and the very prime inclination of nature" ([19], pp. 532, 534). In this same year, John Wesley urged Pitt to discourage suicide by hanging the corpse in chains for all to witness ([29], pp. 139F.).

Horne's work was a reasoned theological defense, but it was a short tract and less substantive than Charles Moore's *A Full Enquiry into the Subject of Suicide*, published in 1790. Moore was mildly critical of Donne, but reserved an especially stinging rebuke for Hume:

> To combat Donne... is in fact to answer almost all the material arguments that have been used by modern defenders of suicide... He only wished to maintain, that there were reasons, which might make it (if not sometimes a meritorious, yet) an indifferent action... It will be found however in the examination of the book itself, that his arguments tend to overthrow all the principles and laws on which the general guilt of

suicide is established; and that therefore if valid, they open the way to a much more frequent commission of the crime than Donne himself thinks allowable. [Hume should have known that] a philosopher and an oyster have each their respective stations. The oyster fulfils the law of his existence (whatever it be) as long as he lives, and at length dies (though not through suicide) for the nourishment of the philosopher. The philosopher leads a life of learned ease, which he employs in ingeniously arguing (or attempting to argue) away the first instinctive principles of nature, reason, faith, religion ([30], pp. 6, 9, 54).

In the most exhaustive and best argued anti-suicide tract of the 18th century, Moore judged Hume a pernicious and destructive member of British society. Hume would ultimately win the battle for the British mind, beginning perhaps in the Parliamentary debates of 1823, but in 1790 there was little doubt that Moore's theses were generally viewed as truth and Hume's as those of a Scottish infidel. And even when the tradition of burial at the crossroads was ended by statute in 1824, burial was still to be at night and without religious rites.

V. KANT

It is almost universally agreed that the most accomplished philosopher to emerge from German culture in the Age of Reason was Immanuel Kant, who was stimulated to some of his most notable theories in metaphysics and epistemology by Hume's writings. Whether he had read Hume or Donne on suicide is not known, but Kant's work on suicide is typical of his rejection of what he saw as Hume's skepticism. He also vigorously rejects the assumption found in Seneca and Hume that happiness or living well is a goal of life relevant to a decision for suicide. Kant lays emphasis instead on who *merits* happiness based on moral performance and on our lack of a moral license to take our own lives.

As with Hume, we possess only short fragmentary work by Kant specifically devoted to the topic of suicide: a lecture joined with two pages on the duty to preserve one's life. However, he used suicide as one of the primary examples of a wrongful act in the course of developing his general moral philosophy, which contains five brief discussions of suicide in his three main writings. Kant uses an original and ornate philosophical framework apart from which his views on suicide lack intelligibility. The interpretation of his works requires the context of his emphasis on performance of duty, where he insists that we must act not only in accordance with duty but for the sake of duty. That is, a moral act requires a recognition by the agent of the act as resting on and intending to fulfill duty. It is not good enough that one performs the morally correct action in the knowledge that it is correct, because one could be acting on self-interested reasons. Like any act, suicide could be morally justified only if it fulfilled duty, or at least was not prohibited by duty. Kant, however, argues that it is prohibited by duties to oneself and to God, and cannot be justified by appeal to rights of autonomy.

Categorical Imperatives and the Prohibition of Suicide

Kant tries to establish the ultimate basis for the validity of moral rules in pure reason, not in intuition, conscience, or utility. He envisions morality as providing a rational framework of principles and rules that constrain and guide everyone, independent of personal goals and preferences. Considerations of public utility and private self-interest are secondary, because the moral worth of an agent's action depends exclusively on the moral acceptability of the rule according to which the person is acting – or, as Kant prefers to say, moral acceptability depends on the rule that determines the agent's will. An action has moral worth only when performed by an agent who possesses what Kant calls a good will; and a person has a good will only if moral duty based on a valid rule is the sole motive for the action.

Kant develops this notion into a fundamental moral law, which he formulates as: "I ought never to act except in such a way that I can also will that my maxim should become a universal law." Kant calls this principle the "categorical imperative". It is categorical because it admits of no exceptions and is absolutely binding. It is imperative because it gives instruction about how one must act. This categorical imperative serves in Kant's system as the criterion of moral correctness for any practical rule on which an agent might act. He gives several examples – all controversial – of moral maxims that are made imperative by this fundamental principle: "Help others in distress," "Do not commit suicide," and "Work to develop your abilities." What makes rules of duty such as "Do not commit suicide" correct, in Kant's view, is their universality. Suicide could not be made universal, he believes, without destroying one or more moral rules presupposed by the action. If suicide violates the rules presupposed by the system, it renders the system "inconsistent" – i.e., having inconsistent rules of operation – and thus is rationally unacceptable.

Kant's categorical imperative is a notoriously obscure and difficult principle, and it is particularly difficult and problematic in the case of suicide.

Here are two of his statements of his major pronouncements on suicide, in which he attempts to apply the categorical imperative:

> A man feels sick of life as the result of a series of misfortunes that has mounted to the point of despair, but he is still so far in possession of his reason as to ask himself whether taking his own life may not be contrary to his duty to himself. He now applies the test "Can the maxim of my action really become a universal law of nature?" His maxim is "From self-love I make it my principle to shorten my life if its continuance threatens more evil than it promises pleasure". The only further question to ask is whether this principle of self-love can become a universal law of nature. It is then seen at once that a system of nature by whose law the very same feeling whose function is to stimulate the furtherance of life should actually destroy life would contradict itself and consequently could not subsist as a system of nature. Hence this maxim cannot possibly hold as a universal law of nature and is therefore entirely opposed to the supreme principle of all duty ([26], p. 59; *Ak.* IV, 53–54). Man cannot renounce his personality so long as he is a subject of duty, hence so long as he lives; and that he should have the moral title to withdraw from all obligation, i.e. freely to act as if he needed no moral title for this action, is a contradiction. To destroy the subject of morality in one's own person is to root out the existence of morality itself from the world, so far as this is in one's power; and yet morality is an end in itself ([28], p. 85; *Ak.* IV, 421–422).

In these and related passages, Kant seems to be arguing as follows: It is impossible to will that suicide from self-interest become universal law without destroying the possibility of willing universal laws. The moral system presumes a body of rules all of which entail a moral agent who acts on the rules, but suicide presumes that it is permissible to act otherwise. Consider what it would mean to universalize or consistently recommend the apparent Stoic rule, "Commit suicide when it works to your advantage." A morality of obligations is inconsistent with this rule of behavior, which would not require persons to do their duty. The act of suicide is thus inconsistent with the rules and practices of morality that it presupposes, producing the inconsistent rules "Do one's duty" and "Escape from duty."

Although this exegesis follows the main lines of Kant's theory of the categorical imperative, the argument is not very plausible and not clearly even his leading argument when discussing "duties to oneself". In this section of his work he demands a consistency between the objectives of an actor's maxim and what he calls "purposes of nature". Kant holds that the function of self-love is to aim at life and not at death, and thus the living person considering suicide contradicts the purpose of life by purposing death. "Because all nature seeks its own preservation," "man sinks lower than the beasts by such an act," abasing humanity that was "entrusted to man for its preservation" ([24], p. 150; [28], p. 85). This is the Kantian equivalent of the natural law tradition emanating from Aquinas, although the source of teleology in moral nature is narrowed to the autonomous subject in Kant.

H.J. Paton has argued that "This is the weakest of Kant's arguments... Unless we have an exaggerated idea of the perfection of teleology in nature, unless indeed we commit ourselves to some theory of the working of divine Providence, this argument can carry little conviction except to those already convinced" ([33], p. 154).[15] Paton correctly maintains that Kant begs the question by presupposing the truth of his conclusion. Nevertheless, there is more to Kant's argument than meets the eye. He seems to hold that self-love *could not* be promoted by an act of suicide that alleviates severe pain or some other condition not worth enduring. His point appears to be *not* that there can be no *empirical* circumstances in which the alleviation of pain would be welcomed by an agent; of course there can be, as we see in the examples of suicides committed every day. Rather, his point is that the proper end of a rational being is self-preservation. To destroy one's ability to act is in contradiction with objectives of free rational agents as such, and so the suicide is in contradiction with his or her very nature.

Kant's vision of a moral system of nature and proper action is not based on any empirical experience or attitudes (see the discussion of "heteronomy" below) but rather on our *nature* as rational autonomous beings. His moral realm is grounded in rational human nature as such, not in the human experiences of individuals. When Kant says that suicide would destroy the system of nature, he does not mean *empirical* nature but rather *rational* nature. His contention is that "*the* autonomous person" constitutes that nature, and the whole order of nature is contingent on the existence of *that* "person". The bottom line seems to be that suicide is immoral because it attacks the moral order itself: Suicide makes self-love the end of action, and the moral agent's life becomes a means to this end in abandonment of its proper ends.

Kant is convinced that a suicide is not only a violation of one's duty to oneself but to God as well. As a "rebel against God" the suicide "leaves the post assigned him", as a "sentinel on earth" and "violates a holy trust." Building on these metaphors, he invokes others: "As soon as we examine suicide from the standpont of religion we immediately see it in its true light... God is our owner; we are His property" ([28], p. 85; [24], pp. 151–154). These arguments are no stronger than St. Thomas' similar ones, and again need the prop of both theology and a vision of nature's teleology to support the moral claims. (However, for Kant, unlike St. Thomas, the moral order is constituted not by God's command, but by the order introduced by rationality.)

Paton contends that, "The argument would be more plausible if we were to maintain that to commit suicide only because life offered more pain than pleasure is at variance with the function of reason as aiming at absolute good; for it is to withdraw oneself in the interests of comfort from the duty of leading the moral life" ([33], p. 154). True, this is a better argument, but not by much; and it is too weak for Kant because it leaves the opportunity for justifying suicide that he apparently wishes to preclude by his emphasis on the grounding of moral correctness in human nature rather than in contingent human experiences. In his *Lectures on Ethics*, where his major fragment on the topic of suicide is found, he argues that Cato's suicide "to prevent himself from falling into Caesar's hands" is "the only example which has given the world the opportunity of defending suicide. It is the only example of its kind and there has been no similar case since" (p. 149). In any event, Paton's substitute argument is not found in Kant's writings.

The most intriguing moral questions about suicide raised by Kant are never answered, because they do not admit of clear answers from his principles. In a series of "casuistical questions" cast as an exercise in seeking the truth where no dogmatics is possible, he asks whether suicides not done from (empirical) self-love can be justified. In one case he considers whether a man "does wrong" if he is bitten by a rabid dog, contracts incurable rabies, and kills himself in order to avoid harming others when delirious. Kant also asks whether a great king acts wrongly by poisoning himself after a losing battle in order that his country not be disadvantageously ransomed ([28], pp. 86–87).

Kant never resolves these questions about moral motives in conflict, and we can only speculate at his answers. On the one hand, he apparently believes that such acts are wrong, on the anti-suicide grounds discussed previously. On the other hand, most of the motives to suicide that he cites in his casuistical questions seem to be morally praiseworthy, and if Cato's suicide is justified and these cases relevantly resemble Cato's act – as they seem to – then Kant must sanction these acts as well. This lack of resolution indicates that Kant's categorical imperative is relatively powerless to resolve such questions, but also that Kant may never have come to grips with the objection that the more one is disabled from the capacity to perform duties by imprisonment, injury or terminal illness, and the like, and the more one is not in a position to act in the interests of others, the less one has a duty to self, others, or God to preserve life. This is Hume's view, of course, and Kant does not seriously entertain or counter it.

Kant is clear, however, that allowing oneself to be killed is not the moral equivalent of killing oneself, and that the former is easier to justify than the latter ([24], p. 152). Kant suggests that one has a *duty* to allow oneself to be killed in certain circumstances, although he is unclear on the nature of the distinction between killing and allowing to die, and even offers enigmatic examples such as the following that challenge the very distinction on which his arguments rest:

> If a man cannot preserve his life except by dishonouring his humanity, he ought rather to sacrifice it; ... If, for instance, a woman cannot preserve her life any longer except by surrendering her person to the will of another, she is bound to give up her life rather than dishonour humanity in her own person, which is what she would be doing in giving herself up as a thing to the will of another. The preservation of one's life is, therefore, not the highest duty... A higher duty [can come] into play and commands me to sacrifice my life (*ibid.*, pp. 156–157).

Presumably, Kant is thinking only of cases in which one suffers death because of a failure to comply with *immoral* commands of others that would morally degrade the person if coerced "consent" were obtained. However, it is hard to understand in the rape case how the distinction between killing and being killed could be meaningfully applied, how consent could be meaningfully given in circumstances of coercion, or why it is worse to kill oneself than to be killed by refusing the advances of the would-be rapist. Moreover, Kant avoids treating cases in which a refusal will lead to *degradation* (by rape, say) but *not to death*. In short, Kant does not probe deeply into his own qualifier that a higher duty can command a person to sacrifice his life; indeed, he fails to specify the scope of the word "sacrifice". While he clearly condemns suicide in the tragic cases of *physical* degradation of the sort so important to Hume and the Stoics, he has severe problems about how to handle the justifiability of tragic cases of *moral* degradation. Once again he does not seem to have appreciated or met the Humean challenge.

The Limits of Respect for Autonomy

Kant has been cited by certain patrons of liberty for his eloquent defenses of human autonomy and the right to act autonomously. The myth that Kant defends *individual* choice in moral matters such as suicide has grown up around his famous doctrine that persons must always treat each other as autonomous ends and never as means only. This often misunderstood thesis is not a categorical demand that we never treat others as means to the ends of personal or public goals – as, for example, employers and government officials treat persons daily. His imperative demands only that such persons be treated with the respect and moral dignity to which every person is entitled at all times, including those times when they are used primarily as means to the ends of others. Kant's fundamental thesis is that to treat persons merely as means, strictly speaking, is to disregard their personhood by exploiting or otherwise using them without regard to their existence as autonomous agents. He held it to be categorical that we must never treat another as a means to our ends without also recognizing that he exists as an autonomous end.[16]

Interpreters have sometimes distorted this Kantian thesis by presenting Kant as a defender of the individualistic principle that every individual has the right to legislate rules and make personal decisions, and hence to be "one's own" person through self-rule. In this conception, "autonomy" is to be understood as the capacity of persons to reflect critically on and take responsibility for the persons they choose to be. What makes a life that person's own is precisely that it is shaped by his choices, and this principle seems to leave the door open for the choice of suicide as well as for a condemnation of anyone who would interfere with an act of rational suicide.

This conception of autonomy, however, is *not* Kant's. It is attributed to Kant in the literature only through uncritical conflation of the broad family of ideas associated with autonomy in contemporary moral and legal philosophy: rights of individual liberty and privacy, free choice, choosing for oneself, being one's own person, creating one's moral position, accepting ultimate responsibility for one's moral views, and the like. The vernacular of autonomy in contemporary moral theory has been erected on individualistic rather than Kantian foundations, but there has been a desire to harness Kant's celebrated moral theory to the view that a principle of respect for autonomy requires that the reasons for an autonomous person's actions be respected and that the actions themselves not be prevented or otherwise restrained (unless harmful to others or otherwise immoral). Suicide is clearly justified on such a principle of respect for autonomy in *Hume's* philosophy. Kant, however, is opposed to all libertine, individualist, and anarchist conceptions of autonomy, and it is critical to understand how and why he is so opposed in order to understand his views on suicide.

It is true that Kant's theory of autonomy can be appropriately explicated in terms of *self-legislation:* If a person freely determines the principles under whose direction he will act, the person is a law-giver to himself, and thus is autonomous. However, this spare interpretation can be misleading, because it states only a necessary, not a sufficient condition of autonomy. For Kant, autonomy of the will is exclusively moral;[17] autonomy is present if and only if one knowingly governs oneself in accordance with *universally valid* moral principles that survive application of the categorical imperative ([27], p. 51, *Ak.* IV, p. 432). That is, only if a person behaves according to rationally self-legislated moral rules specifying duties does Kant consider that person to act autonomously or to have an autonomous will. "Will", as a kind of causality belonging to those who are rational, is the power of acting causally in accordance with principles, and willing is acting from this power (*ibid.*, pp. 29, 45, 64; *Ak.* IV, pp. 412, 427, 446).[18]

Autonomous actions thus must be premised on universally valid moral reasons, and any act that conflicts with objective morality is forbidden (*ibid.*, p. 58, *Ak.* IV, p. 439). The relation between autonomy and moral universality Kant locates in pure rationality: To will rationally is to will exactly what any creature distanced from personal preference would will. Such a person is, functionally speaking, an agent deprived of all knowledge of his own particular person. Thus, for Kant what morality requires is determined by what we would do "if reason completely determined the will" ([25], p. 118; *Ak.* V, p. 20). Kant insists that this subjection to universal rules is consistent with *moral* autonomy. Because the person accepts and wills the principle determining the action, the person is complying with *his own rule* – not with a foreign influence or source of obligation. The person's dignity – indeed, "sublimity" – comes not from subjection to the law but rather from *being* the lawmaker, i.e., from being autonomous ([27], p. 58; *Ak.* IV, pp. 439–440). This view is sometimes called the "paradox of Kantian ethics", and it is indeed a paradoxical account of autonomy: Kant holds that, morally speaking, the self is entirely governed by law, and yet the law is completely dependent on and derivative from the self.

In a famous rebuttal, Henry Sidgwick objected to Kant that the principles of the scoundrel and the principles of the saint could both be selected autonomously. This objection springs from a misunderstanding. John Rawls has recently responded, as a Kantian, that although a *free* self *could* choose as a scoundrel would, such choice is inconsistent with the choices that free and equal rational beings expressing their nature as such would make. It is not only the liberty to choose that characterizes autonomous choice; such choice must conform to what free and rational beings would select as expressing "their autonomous nature as rational". The autonomous agent, then, wills exclusively from rational principles that are valid for everyone, and actions from this ideal "express most fully what we are or can be" as agents independent of "the contingencies of nature and society" ([34], pp. 252, 256; *cf.* [35], pp. 97ff.).

Any philosophy in which a right to individual autonomy can legitimately outweigh the dictates of objective moral principles of duty is alien to Kant's moral theory. The freedom to form personal moral opinions is an absurdity as such, and any action from personal freedom that overrides the dictates of morality on grounds of the conscientious judgment of a free person is a perversion of morality that deserves condemnation. Suicide simply is not within the boundaries of that which is morally permissible in Kant's estimation. Any "conscientious" agent who determines that respect for individuality justifies the election of suicide has no moral grounds on which to stand in the claim that such a judgment should be respected. As Rawls argues, without an adequate framework of moral objectivity, an individualistic conception of "autonomy is likely to lead to a mere collision of self-righteous wills" ([34], p. 519). Naturally, Kantians do not preclude a theory of valid individual choice in contexts of non-moral disagreement, but suicide does not fall into the non-moral realm.

Heteronomy and Governance of the Person

Kant contrasts autonomy with heteronomy of the determination of the will by persons or conditions other than oneself. If reason is enslaved to passion or to a will controlled by desire, the person is heteronomous rather than autonomous. Legislation by desire or inclination renders the person under the control of affective or sensuous impulses – or at least directs the person by principles not his own. This heteronomy is not mere influence for Kant; it is causal necessitation. Under the heading of heteronomy, Kant includes any determinative control over the will – internal or external – except determinations of the rational will by valid moral principles. Any principle not based on autonomy of the will cannot form the will's *own* law, and Kant thus regards acting from desire, fear, impulse, and habit as no less heteronomous than actions manipulated or coerced by others. A person's desire is of course owned by that person, but it is not autonomous because of Kant's narrow conception of both the self (as that which is common to all rational agents) and autonomy. Non-rational willing from impulse and desire arises from the affective dimensions of human nature and cannot make moral law ([27], pp. 59–63; *Ak.* IV, pp. 441–444). Because Kant sees suicide as a motive to action involving subservience to a source of legislation not one's own, this argument supplies a reason for rejecting the very possibility of an *autonomous* suicide.

In his later writings (expecially his *Religion*), Kant argues that human choice can be free even if in conformity with universal moral laws. That is, a will can be free – independent of causal influences – without being autonomous. For example, evil actions can be freely willed and could have been otherwise (cf. [39], [37], esp. pp. lxxxv, ciii–cxi, cxxvii–cxxviii, and [38]). Will as free choice may act autonomously or heteronomously (i.e., may conform to the moral law or may be determined by a personal goal), but in either case may be freely self-determined. These distinctions are crucial for showing how one can be responsible for and blamed for one's moral mistakes. If one acts freely against the dictates of morality, one is no less responsible than for one's autonomous acts. Only if he accepts this premise can Kant criticize those who fail to conform to their moral obligations, and this is precisely the basis for his criticisms of those who succumb to the motivation of suicide.

V. CONCLUSION

I have argued that, among the corpus of writings on suicide that we today think of as having been historically influential in the Age of Reason, David Hume's writings alone stand as a libertine, secular contribution. However, many premises and arguments are shared by thinkers during this period merely because they carried the argument furthest in defense of liberty. For example, Hume, Donne, and Kant all agree that a suicidal action may be cowardly and even morally wrong. Hume holds that moral duties can be overridden by prudential interests, but he agrees with Kant that morality can triumph over self-advantage in some circumstances of suicide. Because all interests are considered, the overriding obligation may be to abstain from suicide. It would take a more sweeping defense of liberty than Hume's to argue that any autonomous suicide is justified, no matter the conditions under which the autonomy is exercised.

One problem with the discussion of suicide found in St. Thomas and other thinkers who set the agenda for reflection in the Age of Reason is that they failed to distinguish the wrongness of an action from the moral excusability and even praiseworthiness of the intention that underlies that same action. Donne and Hume found this weakness to be a major flaw in the traditional Christian viewpoint. They tried to show that we need to introduce distinctions based on intentions and mitigating circumstances and that if we do, a suicide may not be a person who denies the value or meaning of life, even of his own life. Although some suicides view life as not worth living, others sacrifice their lives for some surpassing purpose, thus making the value of their lives subordinate to the value of another cause. These pleas for caution in the moral assessment of a suicide's intention and action ultimately proved to be the most powerful premise in the overthrow of the traditional Thomistic arguments against which they were directed.

NOTES

[1] A useful discussion of "Criteria for Rational Suicide" is found in [3], pp. 132–153. See also [8] and [18]. On the role of the concepts of manipulation and coercion, see [14], pp. 7, 10.

[2] Contrast Aquinas' discussion of the case of Samson: "... Not even Samson is to be excused... except the Holy Ghost... had secretly commanded him to do this" (*Summa Theologica* 2–4, 64, 5, *ad* 4). *Cf., ibid., ad* 6.

[3] St. Augustine applied the excuse not only to Samson but also to martyred women who cast themselves into rivers in order to avoid being raped ("menaced with outrage") by pagans. Augustine maintained that God commanded them so to act; however, he cautioned that they should "be very sure that the divine command has been signified." Book I, Art. 26, p. 31.

[4] This view is embraced by Spinoza (*Ethics,* part IV, prop. xx, n.) and probably is the view implicit in Freud's distinction between a life instinct and a death instinct. Aquinas, perhaps inconsistently, expresses a closely related thesis in *ST* 2–1, 6, 5.

[5] For this interpretation I have relied on passages in Aquinas other than those concerned exclusively with suicide. In particular, see *ST* 1–2, 90–97, and 2–2, 94.

[6] The principle of double effect allows him this conclusion. Cf., Aquinas' Reply to Objection 4 in Article 5 and Reply to Objection 1 in Article 6 for his excusing condition. Article 7 in Question 64 contains a rough formulation of the principle of double effect. This principle may be formulated as follows: Whenever from a moral action there occur two effects, one good and the other evil, it is morally permissible to perform the action and to permit the evil if and only if:

1. The intention is to bring about the good effect and not to bring about the evil effect (which is merely foreseen).
2. The action intended must be truly good or at least not evil.
3. The good effect must bring at least as much good into the world as the evil effect brings evil into the world.

It is sometimes said that a fourth condition must obtain: the evil effect must not be the cause of or means to the good effect. Whether this condition is a necessary condition of the principle is a substantive moral controversy into which we cannot delve here. It does not seem to be found in Aquinas' text. The principle of double effect itself is only implicit in Aquinas' work (especially in the sections on killing in self-defense) and it was not explicitly formulated for centuries after his death.

[7] *Biathanatos* was licensed on September 20, 1644, and entered in a register on September 25, 1646, then published with an undated titlepage by Henry Sayle. All known presentation copies date at the earliest from 1647.

[8] For the relationship of Donne's organizational structure to Aquinas', see [12], pp. xlvii–xlix.

[9] This essay was originally printed for but not published in Hume's 1757 *Dissertations;* it was first published in English in 1777, as *Two Essays* – an anonymous and unauthorized edition. The essay had been published surreptitiously in French in 1770. I shall refer to the pagination found in [23]. A more standard, but less available, edition is David Hume, "Of suicide" [22].

[10] Sprott has shown, in outline form, that the bulk of Hume's arguments had previously appeared in the work of others: [40], esp. pp. 132f.

[11] Hume imagines still other cases, such as that of a prisoner condemned to die and the presumably historical case of Strozi of Florence. But these cases do not constitute types of cases relevantly different from the third case.

[12] Hume's exact words are: "What is the meaning of that principle, that a [suicide]... has incurred the indignation of his Creator by encroaching on the office of divine providence, and disturbing the order of the universe?" [23], p. 154.

[13] I have argued in detail for this interpretation in [5].

[14] Horne's work, as cited here, includes his "Letters on Infidelity, to which is prefixed a Letter to Dr. Adam Smith", "Remarks on the Apology for Mr. Hume's Life", and "Examination of the Essay on Suicide", as collected in [19].

[15] See also [17], p. 67, and, for a reply to Paton, [32], pp. 85ff.

[16] Alan Donagan's formulation of Kant's principle is accurate and to the point: "No rational being should ever be used merely as a means; always, even when he is being used as a means, he must at the same time be treated as an end" [11], p. 229.

[17] Theories of autonomy from the moral point of view are not original with Kant. Theories were perhaps first developed in the medieval period. See [43], p. 127.

[18] Caution is needed about the word "action". Kant seems to allow an act to be willed where action in the ordinary sense (performance) is prevented.

[19] Here Rawls seems to mean individual rather than moral autonomy.

BIBLIOGRAPHY

[1] Aquinas, St. Thomas: 1948, *Summa Theologica,* trans. by English Dominican Fathers, Benziger Brothers, New York.

[2] Augustine of Hippo, St.: 1950, *The City of God,* trans. by W.J. Oates, Random House, New York, Vol. 1.

[3} Battin, M.P.: 1982, *Ethical Issues in Suicide,* Prentice-Hall: Englewood Cliffs, N.J.

[4] Battin, M.D., and Mayo, D.J. (eds.): 1980, *Suicide: The Philosophical Issues,* St. Martin's Press, New York.

[5] Beauchamp, T.L.: 1976, 'An Analysis of Hume's Essay, "On Suicide"', *The Review of Metaphysics* 30, 73–95.

[6] Beauchamp, T.L.: 1986, 'Suicide', in T. Reagan (ed.), *Matters of Life and Death*, 2nd ed., Random House, New York, pp. 77–124.
[7] Blackstone, W.: 1777, *Commentaries*, 4th ed., Clarendon Press, Oxford.
[8] Brandt, R.B.: 1975, 'The Morality and Rationality of Suicide', in S. Perlin (ed.), *A Handbook for the Study of Suicide*, Oxford University Press, New York.
[9] Clarke, S.: 1706, *A Discourse Concerning the Being and Attributes of God, the Obligations of Natural Religion and the Trust and Certainty of the Christian Revelations, in Opposition to Hobbes, Spinoza, the author of The Oracles of Reason, and other Deniers of Natural and Revealed Religion*, James Knapton, London. Reprinted in D.D. Raphael (ed.): 1969, *British Moralists, 1650–1800*, 2 vols., Clarendon Press, Oxford.
[10] Colie, R.: 1966, *Paradoxia Epidemica*, Princeton University Press, Princeton.
[11] Donagan, A.: 1977, *The Theory of Morality*, University of Chicago Press, Chicago.
[12] Donne, J.: 1982, *John Donne's Biathanatos*, Battin, M. and Rudick, M. (eds.), Garland Publishing, New York.
[13] Donne, J.: 1984, *Biathanatos*, E.W. Sullivan (ed.), University of Delaware Press, Newark, Delaware.
[14] Faden, R., and Beauchamp, T.L.: 1986, *A History and Theory of Informal Consent*, Oxford University Press, New York.
[15] Fedden, H.R.: 1972, *Suicide: A Social and Historical Study*, Benjamin Blom Inc., New York.
[16] Given, J.B.: 1977, *Society and Homicide in Thirteenth-Century England*, Stanford University Press, Stanford.
[17] Glass, R.: 1971, 'The Contradictions in Kant's Examples', *Philosophical Studies* 22, 65–70.
[18] Hauerwas, S.: 1981, 'Rational Suicide and Reasons for Living', in M. Basson (ed.), *Rights and Responsibilities in Modern Medicine*, Alan R. Lisse, Inc., New York, vol. 2., pp. 185–189.
[19] Horne, G.: 1846, *The Works of The Right Reverend George Horne, D.D.*, Henry M. Onderdonk, & Co., New York, vol.2, pp. 507–537.
[20] Hughes, R.E.: 1968, *The Progress of the Soul: The Interior Career of John Donne*, William Morrow, New York.
[21] Hume, D.: 1932, *The Letters of David Hume*, J.Y.T. Greig (ed.), Clarendon Press, Oxford, vol. 2.
[22] Hume, D.: 1963, 'Of Suicide', in *Essays: Moral, Political and Literary*, Oxford University Press, Oxford, pp. 585–596.
[23] Hume, D.: 1965, 'On Suicide', in J. Lenz (ed.), *Of the Standards of Taste and Other Essays*, Bobbs-Merrill, Indianapolis.
[24] Kant, I.: 1930, *Lectures on Ethics*, L. Infield (ed.), The Century Company, New York.
[25] Kant, I.: 1956, *Critique of Practical Reason*, trans. by L.W. Beck, Bobbs-Merrill, Indianapolis.
[26] Kant, I.: 1956, *Groundwork of the Metaphysic of Morals*, trans. by A.J. Paton, Harper and Row, New York.
[27] Kant, I.: 1959, *Foundations of The Metaphysics of Morals*, trans. by L.W. Beck, Bobbs-Merrill, Indianapolis.

[28] Kant, I.: 1964, 'The Doctrine of Virtue', *The Metaphysics of Morals*, part II, trans. by M.J. Gregor, University of Pennsylvania Press, Philadelphia.
[29] Lecky, W.E.H.: 1904, *History of England*, vol. 3, Longmans, Green, London.
[30] Moore, C.: 1790, *A Full Enquiry into the Subject of Suicide*, vol. 2., John Hatchard, London.
[31] Mossner, E.C.: 1950, 'Hume's *Four Dissertations:* An Essay in Biography and Bibliography', *Modern Philology*, 48 (August 1950), 37–57.
[32] Novak, D.: 1975, *Suicide and Morality*, Scholar's Studies Press, New York.
[33] Paton, H.J.: 1967, *The Categorical Imperative: A Study in Kant's Moral Philosophy*, 6th ed., Hutchinson & Co., London.
[34] Rawls, J.: 1971, *A Theory of Justice*, Harvard University Press, Cambridge.
[35] Rawls, J.: 1975, 'A Kantian Conception of Equality', *Cambridge Review* (February), 97.
[36] St. John-Stevas, N.: 1961, *Life, Death and The Law*, Indiana University Press, Bloomington.
[37] Silber, J.: 1960, 'The Ethical Significance of Kant's *Religion*', in I. Kant, *Religion Within the Limits of Reason Alone*, Harper Torchbooks, New York, pp.
[38] Silber, J.: 1962, 'The Importance of the Highest Good in Kant's Ethics', *Ethics* 73, 179–197.
[39] Silber, J.: 1974, 'Procedural Formalism in Kant's Ethics', *The Review of Metaphysics* 28, 197–236.
[40] Sprott, S.E.: 1961, *The English Debate on Suicide*, Open Court, La Salle, Illinois.
[41] Sym, J.: 1637, *Life's Preservation*, London.
[42] Taylor, Jeremy: 1660: *Ductor Dubitandum, or the Rule of Conscience*, 2 vols., London.
[43] Ullman, W.: 1967, *The Individual and the Society in the Middle Ages*, Methuen, London.
[44] Watts, I.: 1810, 'A Defence Against the Temptation to Self-Murder' (1726), in *The Works of the Reverend and Learned Isaac Watts, D.D.*, vol. 2, Clarendon Press, Oxford.

Georgetown University,
Washington, D.C., U.S.A.

JOSEPH BOYLE

SANCTITY OF LIFE AND SUICIDE: TENSIONS AND DEVELOPMENTS WITHIN COMMON MORALITY

I. INTRODUCTION: SANCTITY OF LIFE AND COMMON MORALITY

Among the positions taken on the various disputed questions related to the ethics of killing, one has come to be called the "sanctity-of-life" view. What is designated by this expression is not a carefully articulated philosophical position, but a family of connected views and attitudes, not all of which would be accepted by most who would consider themselves committed to the sanctity of human life.

What most obviously unites these views and attitudes is that those who hold them tend to reject as immoral virtually all killing by private persons. The major exception is killing in self-defense. Publicly authorized killing, whether in war, or in punishing convicted criminals, or in police activity, is usually held to be in a different category: such killing is thought by some to be required by a proper regard for the sanctity of innocent life, and by most to be compatible with the sanctity of life. Thus, what seems to define the sanctity-of-life view is opposition to abortion, to the various forms of euthanasia, and to suicide. This opposition is frequently expressed in absolute, exceptionless prohibitions of actions of these kinds, although the absolute prohibitions are usually formulated so as to refer to classes of actions less extensive than the ordinary usage of terms like "abortion" and "suicide" suggests. Thus, the absolute prohibition of abortion by those holding sanctity-of-life views is usually understood as a prohibition of "direct" or intentional abortion.

The sanctity-of-life position, therefore, does not regard the ethics of killing as simply an application of justice to the questions of human life. It is not simply a "right-to-life" position, if the right is taken as a person's claim against others not to take his life. For the sanctity, or sacredness, or dignity of human life, according to this view, provides reason why one should not kill an innocent person, even if that killing should violate no norm of fairness or, for that matter, any other relevant moral norm.

Baruch A. Brody (ed.), Suicide and Euthanasia, pp. 221–250.
© 1989 *Kluwer Academic Publishers*.

However, the popular tendency to identify the sanctity-of-life and the right-to-life positions is suggestive: the sanctity of human life is usually taken not only to have implications for the decisions of those who regard it as true, but also to be a proper basis for legislation on such matters as abortion, euthanasia, and suicide, including voluntary euthanasia and assisted suicide.

As the word "sanctity" and its common synonym "sacredness" suggest, this position has religious roots. In fact, the position is based rather directly on Christian ethics and, in particular, on the articulation of that ethics developed within Catholicism especially since the time of Thomas Aquinas. However, the philosophers who have articulated and defended versions of a sanctity-of-life position have not regarded their convictions as simply religious, that is, as simply implications of the special duties to God included within their specific religious belief. They have regarded it instead as an implication of the natural law or, in Alan Donagan's apt description, a part of "common morality".

According to Donagan, common morality is a part of the morality articulated originally in the Hebrew and Christian scriptures, and developed within the Jewish and Christian traditions. It is that part of this morality which, according to these traditions themselves, can be known by common experience and rational reflection independent of any religious convictions specific to these traditions. Donagan believes that common morality is limited to duties to self and to other human beings, that is, to the duties indicated on the second tablet of the Mosaic law ([7], pp. 6–9).

Even this rough characterization of the sanctity-of-life view and its sources raises a number of questions: Is there a coherent justification for the main normative conclusions of the sanctity-of-life view – in particular, the conclusion that suicide is always morally impermissible? If there is, is this justification compatible with other moral norms usually thought to be part of common morality? And is this justification really a moral justification, and not an implication of religious conviction?

The purpose of this paper is to consider such questions as these and some of the answers given to them by philosophers who accept common morality, to explore some difficulties these answers generate, and to indicate some developments and theoretical options which may be needed if common morality is to deal with these difficulties.

Common morality does not have the coherence of the work of a single thinker or even a single school of thought. For it is an historically extensive and developing tradition containing a variety of emphases and theoretical formulations, as well as many internal strains and tensions. Therefore, I will take as my starting point the work of one representative thinker within common morality – Thomas Aquinas. With the possible exception of Kant, Aquinas has contributed the most to the philosophical formulation of common morality. Furthermore, his discussion of the ethics of killing introduces sanctity-of-life elements in an unmistakable way. The work of contemporary Catholic philosophers who have tried to develop the sanctity-of-life elements in Aquinas' view into a coherent ethical position, and the challenge to that view contained in the recent work of Alan Donagan, perhaps the most impressive recent attempt to give common morality philosophical articulation, will serve to exhibit the complexities and difficulties to which the attempt to ground a consistent sanctity-of-life ethic in common morality gives rise.

The purpose of this paper, therefore, is not as straightforwardly historical as most of the other contributions to this volume. The purpose is to clarify the issues that arise when one looks closely at a tradition of moral thinking which seems to have something distinctive and important to say about the moral and public-policy questions concerning suicide. Since the public-policy questions concerning suicide raise a number of complex issues distinct from the straightforwardly moral questions, and since common morality's approach to these public-policy questions is at least as complex as its approach to the ethics of killing, I will limit my discussion here to the strictly moral questions.

II. AQUINAS' ARGUMENT AGAINST SUICIDE: A PRELIMINARY STATEMENT

Within the religious sources of common morality there are several general condemnations of murder and specific forms of homicide. Although these prohibitions may appear relevant to the evaluation of suicide, they are not, within the scriptural sources themselves, applied to it. Thus, in its original theological context, the precept of the Decalogue prohibiting murder (Ex. 20: 13; Dt. 5:17) was understood to prohibit any kind of killing of one human being by another except that which was divinely authorized. Killing was believed to be divinely authorized in certain particular cases, and, in general, in the case of those who were forcibly violating just order or who had been legally convicted of having done so. There is no indication that this precept was understood to apply to self-killing. Similarly, the more specific precept "The innocent and the just you shall not put to death" (Ex. 23:7) was articulated originally as a precept for judicial proceedings, and as such forbids treating those not guilty of capital crimes as if they were.

However, many Jews and Christian have regarded the Noachite precept "For your own lifeblood, too, I will demand an accounting" (Gen. 9:4) as prohibiting suicide.

Of course, the opposition to suicide by those who accept common morality has not been based simply on this rather obscure text or on an extension or misreading of the other scriptural prohibitions of homicide. For this opposition seems to arise much more from their understanding of the philosophical and theological reflection which developed and systematized the ethics of killing suggested by the traditional sources. This reflection was not tightly tied to a literal reading of specific texts within the sources, but was based instead on convictions about the entire view of human life presented in the tradition. Since many of these philosophers believed that the ethics of killing was a matter of natural law, a part of common morality rather than religious conviction, the looseness of the fit between their views and the traditional sources should not be surprising.

Aquinas' ethics of suicide, and more generally of killing, is a case in point. He undertook it within a specifically religious and theological context, and he was at considerable pains to accommodate the theological tradition, including the relevant scriptural texts. But his discussion unmistakably contains moral, philosophical elements.

For Aquinas, suicide is taken to be a form of murder, a kind of homicide which is morally flawed in essentially the same ways as other prohibited kinds of killing. Thus he cites as the authority for his condemnation of suicide, without any suggestion that he does not accept it, the following argument of Augustine: "It remains that we understand it to be about man that it is said: You shall not kill. Therefore, neither another nor yourself. Therefore one who kills himself kills nothing other than a man" (*ST* 2–2, 64, 3 *sed contra*).[1]

Aquinas was aware that suicide differed in important ways from other forms of homicide. He considered the objection that homicide is a sin because of its injustice, and suicide cannot be unjust. He responded by denying that suicide is unjust in relation to God and to the community, but went on to argue: "Homicide is a sin, not only because it is contrary to justice but because it is contrary to the charity which a man should have toward himself. In this respect suicide is a sin in relation to oneself" (*ST* 1–2, 64, 3 *ad* 1). Thus, Aquinas seems to have thought that what morally flaws suicide, questions of justice aside, is a feature of this kind of killing found, though in slightly different form, in all kinds of prohibited homicide. The charity owed oneself is also owed to any human being.

Aquinas provided three reasons why it is illicit for anyone to kill himself. One of these was that suicide violates God's dominion over human life. His second reason was that suicide is an injury to the community of which the suicide is a member. A third reason (listed first by Aquinas) was that suicide is contrary to the natural love and charity a person should have for himself (*ST* 2–2, 64, 3).

No doubt, the first of these reasons is the main reason why Christians and Jews have opposed suicide. As Donagan notes, however, this seems to be a religious reason and not a moral one ([1], p. 77). Perhaps, however, some adherents of common morality would have regarded this consideration as properly moral. For, many within that tradition would not accept Donagan's restriction of common morality to duties to other humans and to oneself, but would hold instead that there are natural moral obligations to God – obligations which can be known independently of revelation and particular religious conviction. Such obligations, it seems, would presuppose natural theology. There is every suggestion that Aquinas believed that there are such natural obligations to God.[2]

However, even granting that there are natural moral obligations to God, and that natural theology is sufficient for knowing something about God's relation to human beings, it is difficult to see how the claim about God's dominion over human life could be justified independently of revelation and specific religious belief. This claim, understood as a premise in an argument against suicide, seems to indicate a more specific relation between God and human beings than can be read off the ontological structure of the relationship between creature and creator, or off the properties of God as knowable by natural theology. Further, this argument, even if it could be developed as a moral argument, can hardly illuminate the features of human life which give it the sanctity or dignity that would exclude suicide.

The argument that suicide is an injury to the community was not developed by Aquinas. No doubt, this is an important moral consideration in many deliberations about suicide, and it has some bearing on public policy concerning suicide. But it does not seem that considerations of this general kind could justify anything like the general prohibition of suicide which Aquinas has in mind, and which sanctity-of-life views generally support. For surely some people could correctly judge that their continued existence would be more of a burden than a benefit to the various groups which might have claims on them. And in some cases people could correctly judge that they have no further duties to any of the communities to which they belong. Further, like the argument from God's dominion over human life, this argument from the requirements of one's obligations in justice to other human beings does not reveal that feature of human life which demands that it be immune from destruction.

It seems, then, that if there is a rational, non-theological argument for anything like an absolute prohibition of suicide, based on considerations about the dignity or sanctity of human life, it is to be found in the direction of Aquinas' argument that suicide violates the natural love one should have for one's own existence.[3]

Aquinas' argument on this point is as follows: "[Suicide is altogether illicit...] because anything whatsoever naturally loves itself, and it pertains to this that everything naturally conserves itself in existence and resists insofar as it can its corruption. And therefore, the fact that someone kills oneself is contrary to natural inclination and contrary to charity, by which everyone ought to love oneself." This opposition to natural inclination is plainly what Aquinas took to be the reason why suicide is contrary to the natural law.

The essential inference seems to be:

1. Any action contrary to a person's natural inclinations is a violation of the person's obligation to love himself.
2. Suicide is an action contrary to the natural inclination to preserve life.
3. Suicide is an action that violates a person's obligation to love himself.

III. DEVELOPING THE THOMISTIC ARGUMENT

(1) is the most obviously troublesome premise in this argument. For suicide is clearly opposed to any inclination one might have to preserve oneself in being, and staying alive is surely one of the deepest seated and most abiding interests people have. But (1) regards natural inclinations as morally significant. How can natural inclinations ground moral obligations?

The answer Aquinas gives elsewhere is that the objects of the natural inclinations of human beings are immediately known by practical reason to be the fundamental goods of human nature. As such, they are starting points of practical reasoning. Human life is one of these goods (*ST* 1–2, 94, 2; *cf.* [9]). So, what functions in moral arguments, as in practical thinking generally, are not the inclinations themselves but the basic human goods to which they point. Therefore, (1) can be reformulated as (1a): Any action contrary to a person's natural goods is an action that violates a person's obligation to love himself; and (2) can be reformulated as (2a): Suicide is an action contrary to the basic good of human life.[4]

This reformulation does not suggest why the goods of human nature should establish moral obligations. Aquinas clearly thought they do, but his account of why is so central to his approach to morality that it is difficult to sum up briefly. Two points are essential. First, he regarded the goods of human nature as perfections or fulfillments of the various dimensions of the person (*ST* 1–2, 18, 1; *cf. ST* 1–2, 10, 2, *ad* 2). Thus, the actions which instantiate these goods are not things apart from the person, but aspects of the full being of the person. As such, the goods of human nature are components of beatitude. Second, he maintained that the most basic moral principle, from which all others follow, is the twofold love commandment (*ST* 1–2, 100, 3, *ad* 1). Love of human beings plausibly involves a concern for their well being, and this includes a concern for their full being as human persons.

These considerations, however, do not show how a concern for these goods is actually brought to bear in moral reasoning, and here Aquinas himself provides relatively little help. One relevant consideration from his moral writings is that they contain no hint that Aquinas thinks that a rational concern for persons and their full being can be expressed by anything like a consequentialist calculation of the greater good or lesser evil. Thus, it is more reasonable to understand Aquinas as holding that persons and their goods deserve respect, except, perhaps, when it is necessary for authorities to stop criminals' violation of just order or to punish them for it.[5]

Thus, it seems reasonable to understand Aquinas' view that concern for the goods of human nature is morally basic in a somewhat Kantian way: one respects human nature as an end in itself if and only if one cherishes, promotes and respects the goods of human nature, and one fails to respect human nature if one fails to respect all the basic goods of human nature. And an important way in which one fails to respect the goods of human nature is when one acts in ways that damage, impede, or destroy any instance of one of these goods.[6]

Such action is not thoroughly rational but arbitrary, because the good thus subordinated to another has no less claim on our loyalties, is just as much a part of our full being as the other. Acts that damage, impede, or destroy instances of basic goods can be reasonable in the sense that there is a good reason to do them, but they are arbitrary in that they allow concern for one good to motivate actions directed against another irreducible component of the human person. In that precise sense the person is used as a mere means, and not respected as an end in itself.

This explication of why it is rational to regard concern for human goods as basic to morality and to understand that concern as requiring a kind of respect for human goods which excludes actions to damage, destroy or impede instances of them is plausible only if there is some specification of what is to count as an action. For very many, if not all, actions have some negative impact on one or another of the basic human goods.

IV. INTENTION AND HUMAN ACTION

I believe that the specification of the relevant acts – those prohibited because of their harm to basic goods – must be to acts which *intentionally* damage, harm or impede instances of basic human goods. Aquinas could have been more explicit in making this specification in his ethical writings generally and in his discussion of suicide in particular; but he does make this specification in several relevant contexts, and it seems reasonable to suppose that it is presupposed in his argument against suicide.

Shortly after his discussion of suicide, Aquinas considers the question of whether it is licit to kill another in self-defense. In answering this question Aquinas distinguishes between the moral obligations of those having public authorization to defend public order and private persons. The former may intentionally kill malefactors; they can refer the killing in their self-defensive action to the common good. The latter, however, may not intentionally kill even in self-defense. They may undertake defensive action, even action which will predictably lead to the assailant's death, but the assailant's death must be "*praeter intentionem,*" outside the agent's intention in acting. The death, in other words, must be a side effect of the defensive action (*ST* 2–2, 64, 7).[7]

This distinction between what one intends in acting and what is outside the agent's intention is not an *ad hoc* invention of Aquinas to deal with the issue of self-defense. It is a quite general distinction used by him in a variety of contexts including both discussions of particular moral issues and general analyses of human action.[8]

In the discussion of self-defense and elsewhere Aquinas maintains that "moral acts receive their species according to what is intended".[9] By this he seems to mean that what makes a voluntary act the sort of act it is from the moral point of view – what, in other words, is essential to the voluntary act for evaluating it morally – is what the agent intends in acting.

It is not clear that Aquinas holds consistently to this conception of human acts as essentially determined from the moral point of view by what the agent intends in doing them.[10] Nevertheless, it seems reasonable to understand the acts referred to in (1) and (1a) above as acts done with the intention to damage, harm, or impede natural inclinations of human goods. For, as noted above, some specification of the acts in question is called for, and the only general way Aquinas uses to distinguish within a voluntary act between what is essential to the act for purposes of moral evaluation and what is not is by appealing to the distinction between what is intended and what is *praeter intentionem*.[11]

If the textual reasons for understanding Aquinas' premise (1), and my reformulation (1a) to refer only to intentional actions are sound, two further questions arise. First, how does Aquinas understand intention, and draw the line between what is intended and what is outside the agent's intention? And second, how does Aquinas justify attributing to intention the considerable moral significance he gives it?

In response to the first question, it seems to me that, although the line drawn by the distinction is not always easily determined, the principle for drawing it emerges clearly enough from a careful reading of the various things Aquinas says about it. The principle, I believe, is this: one intends the state of affairs one takes as one's goal in acting or sees as the benefit of acting, and one includes within this intention any other state of affairs one is committed to realizing as a way to achieve this benefit. Other states of affairs that are connected with these as causal consequences or as other aspects of one's performance in seeking to realize one's goals are not intended. They are neither the benefit one seeks nor precisely what one is committed to realizing as a means to getting the benefit. They fall outside of the good one aims to realize in acting.[12]

This view of intention is not justified by appeal to ordinary language, although there is some support in ordinary English usage for it ([13], [6]). What sustains this view is an analysis of willing, of volitional initiatives. The idea is that voluntary actions are those undertaken on the basis of the judgment that there is some benefit – something valuable or worthwhile – to be achieved by an action.[13] So the action is undertaken for the sake of that aspect of it in virtue of which it is judged to be good. Other aspects of the action, if known, are willed; they are knowingly brought about, and they might provide reasons for not doing the action. But they are willed in a way different from what is precisely intended or chosen. For such aspects of one's act are not what one is interested in realizing: they lack direction to the end.[14]

In other words, it seems to me that what, according to Aquinas, is intended in acting does not extend to every aspect of one's action considered as a singular event but is limited to those aspects of the act corresponding to one's plan of action as articulated in practical reasoning. This plan includes one's end insofar as this is a benefit, and one's means insofar as they are judged to contribute to the realization of the benefit one seeks. Side effects as such are often considered in practical reasoning, but are not what one seeks to realize. They are frequently a matter of indifference, sometimes factors in spite of which one acts, sometimes grounds for not acting, and sometimes factors hindering the realization of that for the sake of which one acts. The failure of expected side effects to obtain does not flaw one's plans, and in some cases might be a cause for satisfaction ([4], pp. 534–537; [11], pp. 229–249).

This account indicates how Aquinas would distinguish between what is intended in action and what is not; and the ground for the distinction in his theory of practical reason and human motivation. This account does not reveal what justifies the moral significance attributed to this distinction in Aquinas' work and in the subsequent Catholic tradition. Aquinas provides little help on this beyond the suggestions already noted. The following considerations, not based directly on the Thomistic texts, provide a starter for such a justification.

The significance of the distinction between what is intended and what is outside the intention is often misunderstood, and the distinction sometimes abused. The abuse arises by supposing that one can "direct one's intention" away from what one actually wants to achieve and so avoid responsibility for it. But one's intention is not captured by just any description of one's act; it is the resolve to bring about some state of affairs, and once the resolve is set one has no further choice about what one intends.

The misunderstanding arises by supposing that one is not responsible for foreseen and accepted side effects. One plainly is. As noted above, such side effects can play a role in practical thinking. When one acts in spite of the disagreeable side effects one expects, one knowingly brings them about, and that surely is voluntary, even though they are not desired or sought. This is recognized by Aquinas and by the more recent proponents of the double-effect doctrine. For it is clear in Aquinas' discussion of killing in self-defense that the assailant's death being *praeter intentionem* is a necessary but not sufficient condition for undertaking the predictably lethal, defensive action. For example, the force used must be no more than is needed for the defense. Similarly, the doctrine of double effect is always stated with a further condition beyond those designed to guarantee that the bad effect is not intended, a condition usually formulated as requiring that there be a grave, or a proportionately grave reason for doing what has the bad side effect.[15]

As I see it, the significance of the difference is that norms that exclude actions harming instances of basic human goods reasonably apply to what is done intentionally, but cannot be reasonably applied to the acceptance of side effects. Such norms reasonably apply to intentional actions because, according to a theory in which concern for basic human goods is central to morality, we should do what we can to avoid harming instances of these goods, and it is always possible to refrain from intentional actions that do this. Sometimes such refraining will require that one accept as a side effect the failure to achieve some benefit, or a harm to some instance of a good which the action could have prevented. But it is never simply impossible to refrain from intentional harms to basic human goods. So the moral approach requires such refraining if it is volitionally possible, and it is possible.

The possibility is different in the case of accepting side effects. For one cannot act at all without accepting some bad side effects (at the very least foregoing the goods to be achieved by alternative actions). So, there could not be a general moral prohibition against accepting side effects harmful to basic human goods. Even if this impossibility does not obtain in all choice situations, it clearly does obtain in many conflict situations in which no matter what one chooses to do, some harm to human goods is inevitable – in which, for example, some people will die no matter what one chooses to do, but where one's choice will settle who dies.

Thus, although actions can be wrong because of their side effects – if, for example, the bad effects of an otherwise good action fall unfairly on some persons – actions cannot be wrong, according to this moral approach, just because they have side effects harmful to instances of basic human goods (*cf.* [8], pp. 291–294).

If this argument is correct, there are grounds compatible with common morality for attributing to the distinction between what is intended and what is not intended but accepted as a side effect the moral significance which many Catholic moralists since Aquinas have given it, at least with respect to moral considerations directly concerned with basic human goods. Since killing human beings directly concerns a basic human good, it can make a moral difference whether the person's death is intended or is accepted as a side effect. It is not, therefore, either simply arbitrary or a matter of moral intuition that intentionally killing the innocent has been taken within the recent Catholic tradition as a moral absolute that expresses the core of the traditional condemnation of murder.[16]

If I am correct in thinking that Aquinas' premise (1) and my reformulation (1a) should be taken as referring to intentional harms to human goods, then the self-killing to which Aquinas' argument applies is only intentional self-killing. This allows a fuller and more precise reformulation of the Thomist-inspired argument against suicide:

1. One should never act with the intention of damaging, impeding, or destroying an instance of a basic good of human nature.
2. Human life is a basic good of human nature.
3. One should never act with the intention of damaging, destroying, or impeding an instance of human life.
4. Intentional self-killing is acting with the intention of damaging, impeding, or destroying an instance of human life.
5. One should never act with the intention of killing oneself.

This conception of suicide applies to many acts in which a person brings about his own death, but not to all such acts. It also applies to many actions and omissions which would not normally be considered suicide, namely, all those including the agent's intention to end his life. Thus, one who chooses to end one's life to avoid the misery and degradation of a lingering death is doing an act of this kind. So also would be the refusal of medical treatment by one who intended this refusal as the way to carry out his decision to end life. Although many would be in doubt about whether to call this suicide, it is intentional killing, and so as much a case of suicide as administering to oneself a lethal injection. This is not to say, of course, that any choice to withhold life-sustaining treatment is suicidal. One may choose to avoid the expense or other costs of treatment and accept death as a side effect. Nor is it to settle whether such a decision should be legally regarded as suicide.

Of the actions which some might be reluctant to call suicide because of the heroism, nobility, or public spiritedness involved, some are acts of suicide and others not. For example, the action of one who altruistically chooses his own death to prevent social or financial embarrassment to his family is an action of this kind. So also are the acts of prisoners who kill themselves to make sure they cannot divulge important information under torture.

But not all such actions, undertaken with the knowledge that they will lead to death, need be suicide in the sense defined. For in many of these cases one intentionally does what will predictably lead to death in order to direct a harm to oneself rather than others or remove a burden from others, without one's death being precisely what prevents the harm or removes the burden. So, for example, actions like that of Captain Oates need not be suicide, but can be the acceptance of death as a side effect of choices to do other things (e.g., removing oneself from the group), which have death as a predictable result.

There is, of course, considerable dispute about how suicide is to be defined. The definition of suicide that emerges from a consideration of Aquinas' views about the relation between action and intention does not attempt to legislate how the word "suicide" is to be used. The moral evaluation of the kinds of acts in question can be expressed without using the term, or by using the term in its ordinary sense. If Aquinas' argument is sound, and if the appeal to intention is defensible, then no definitional questions are begged by concluding that intentional self-killing is always wrong. The truth of this moral absolute is not put in question by the fact that it prohibits a class of acts which is not co-extensive with those we would in ordinary language call "suicide", nor by the fact that there are borderline cases in which it is not clear whether a given action is in fact an act of intentional self-killing, nor by the fact that from some moral perspectives the lines drawn by this approach seem arbitrary.

These definitional issues do raise a serious question of a rather different sort. For the definition of voluntary acts in terms of what is intended is by no means unanimously accepted within common morality. If Donagan, for example, is correct in thinking that one is responsible for one's actions under any description of the actions of which one is aware, and that actions may be redescribed by including within the description causal consequences of the actions, then the difference between intentional self-killing and accepting death as a side effect of other actions will lack moral significance ([9], pp. 37–52).

The obvious implication of Donagan's alternative definition of voluntary action is that the absolute prohibition of intentional suicide based on the Thomist-inspired definition of action cannot be sustained; it draws a sharp moral line where none can be justified, and so either considerations that make it reasonable to accept one's death as a side effect should also make it reasonable to end one's life intentionally, or the reasons that make it absolutely wrong to kill oneself intentionally should be extended to other acts in which one causes one's death. Faced with these alternatives, it seems to me that anyone within common morality would choose the former: self-sacrificing acts that can be foreseen to cause one's death cannot always be immoral.

I think, however, that the issues raised by Donagan's alternative to Aquinas' "intentional" definition of human action are more wide-ranging and profound than their effect on the Thomistic argument against suicide. For if Donagan is correct, Aquinas' understanding of the role of moral absolutes in practical thinking will have to be greatly limited, if not altogether abandoned.

Donagan holds that there are absolute prohibitions within common morality, at least in the sense that once an action has been discovered to violate a perfect duty, it cannot be justified by the fact that it might be necessary to fulfill an imperfect duty. This ordering of perfect and imperfect duties is Donagan's interpretation of what he calls the "Pauline Principle", which prescribes that one must not do evil that good might come about (*ibid.*, pp. 154–157).

Donagan's understanding of common morality's evaluation of suicide seems to be a perfect duty which, in this way, functions as a moral absolute: no one should take his life at will. To understand how this norm functions in practical reason, it is necessary to understand how Donagan understands "at will". Donagan gives two suggestions as to what this phrase means in the prohibition of suicide. First, he seems to regard taking one's life "at will" to refer to the kind of suicide the Stoics justified for sages: one's quitting life as one pleases. This, Donagan maintains, involves treating one's life as cheap, and that is incompatible with respecting oneself as a rational creature. The second suggestion for understanding Donagan's meaning "at will" in his prohibition of suicide is found in his general, indirect strategy for deriving specific precepts from the basic moral principle. He considers a generally described kind of action which appears incompatible with the first moral principle, then examines exceptions, that is, cases in which activity does really not fail to respect human nature as an end in itself; he then lists the exceptions and concludes that other cases of the suspect activity are done "at will" and so morally excluded (*ibid.*, pp. 72–73).

This general procedure does not reveal a feature of a kind of action sufficient to judge its instances incompatible with moral principle. The class of acts thus defined as done "at will" is determined by considering first the exceptions, and there seems to be no principle of closure to guarantee that all the relevant exceptions have been considered. Unless there is such a principle, one can wonder about any member of the suspect class of actions whether it is being done at will, or in fact is another exception, another case of a generally suspect action that is not really incompatible with respect for rational nature.

The special meaning of "at will" used in the case of suicide has an analogously opened texture. For here it is necessary to know when life is being held cheap, and, except in a few clear cases, like, perhaps, playing

Russian roulette just for the thrill of it, this is hardly clear. For people who deliberate about ending their lives usually have weighty reasons for considering such drastic action and need to discover whether in such circumstances they are really holding life cheap. In many such cases, it is hard to see that there is anything but a verbal difference between saying that a given act of suicide is not one in which one is holding life cheap and saying that the duty not to kill oneself is overridden by considerations about one's welfare.

Moral absolutes like Aquinas' prohibition of suicide play a different, and far more decisive, role in the practical thinking of an upright person than norms generated by either of these approaches of Donagan. For on Aquinas' account, a moral absolute is based on some feature of actions sufficient to guarantee that any action of that kind is incompatible with right reason. And that feature is nothing as vague as "holding one's life cheap". Thus, once an action is discovered to be an instance of the prohibition, one knows that whatever else one might think about the action, it is morally out of the question. Sometimes, of course, there are cases where it is difficult to determine whether or not an action falls under the prohibition. But this casuistry can in principle succeed, and is quite distinct from the unacceptable reasoning that overrides moral absolutes by considerations of welfare.

My point here is that Donagan's alternative to Aquinas' use of intention in defining human action not only undercuts Aquinas' absolute prohibition of intentional suicide by rejecting Thomistic theses about the character of voluntary action, but also commits Donagan to procedures for deriving moral precepts which are quite different from Aquinas', and, I think, from that of many others within common morality. Donangan's procedures allow for far too few moral absolutes which function decisively in practical reasoning in the way Aquinas and others within common morality have supposed, and provides a much less perspicuous line between actions that are compatible with the basic moral principle and those in which evil is done for the sake of good than most in the tradition have held.

Of course, there are other alternatives to Aquinas' definition of action than Donagan's. But exploring them quickly leads beyond the confines of this paper. The point of the preceding pages is only that, in the question of how action is to be defined for purposes of moral evaluation, there lie important issues on which proponents of common morality have yet to agree.

V. IS HUMAN LIFE A BASIC GOOD OF PERSONS?

The issues raised by the use of intention are not the only questions which my reformulation of Aquinas' "sanctity-of-life" argument is likely to raise. Both of the key premises in that argument (propositions which together define a sanctity-of-life position) are disputable within common morality. Premise (1) states a moral principle that excludes acting with the intention of harming instances of basic human goods; Donagan clearly rejects this principle ([7], pp. 65, 224–229), and, as we shall see below, Aquinas cannot consistently accept it. Premise (2) is a statement that human life is a basic good of human persons. It too is likely to be questioned by many who accept common morality. For in this tradition rationality appears to be the prized value, and the good of life is most readily understood as instrumental to the higher goods of the self-conscious person.

In spite of this preference for the higher, more spiritual or mental aspects of human life, which Aquinas shares with the entire tradition, there is little doubt that he held human life to be a basic good of human persons. He maintains that the basic principles of practical reasoning, from which moral precepts are derived, are goods that are ordered according to the order of what he calls "natural inclinations". Among these natural inclinations are those we have in common with all things – the inclination to maintain our existence – and the inclinations we share with other living things – to stay alive and reproduce (*ST* 1–2, 94; 2; 1–2, 10, 1). So human life is, for Aquinas, one of the goods of human nature. As such, it is one of the perfections or ends of human life, and when practically understood to be such, it is a basic principle of practical reason. A good's having these features is what makes it to be a basic human good. Aquinas, however, does not provide an extended defense of his claim that life is a basic human good. It is, however, possible to defend this premise in a way compatible with what Aquinas says about life as an irreducible good of human beings (*cf.* [8], pp. 304–309).

Aquinas' view that human life is an irreducible element of human fulfillment and a basic principle of practical reasoning can be amplified in the following way. One often chooses to do something to protect a human life, one's own or another's, without thinking about any good beyond life itself. Of course, life is important for all the other human goods, and so one can have ulterior reasons to protect life. But what is characteristic of a basic good is not that it *always* provides the ultimate reason for an action bearing upon it, but that it can do so. Knowledge of the truth and friendship are goods that often provide the ultimate reason for choices; but even these goods often are pursued for the sake of their contribution to one another or to other goods.

Human life seems also to be an element of the human being's full-being. It certainly is not extrinsic and instrumental as are the possessions persons use. If persons can be said to *have* their lives, they have them in the way that a whole has one of its parts. Moreover, life seems to be not only intrinsic but essential to persons. A human person is a rational, sentient, *living* body. One cannot lose one's life yet continue to be a human person.

These considerations indicate that human life meets the criteria for basic human goodness, and so is not merely instrumental but a basic human good. They do not provide a strict demonstration that life is a basic good of human persons. Fundamental principles are not susceptible of demonstration. On Aquinas' conception of principles this means that the truth of principles is immediately known without the need of a process of reasoning that shows in the middle term of a syllogism the ground for the connnection between subject and predicate. He did not think that principles were therefore assumptions that could not be defended rationally. Rather, he thought that dialectical argumentation could indirectly support the truth of self-evident principles, by clarifying misunderstandings, dealing with objections, and, more generally, showing how the principle in question fits with other knowledge in a way its denial does not. The remainder of this section is an effort of this kind of dialectical clarification.

One common misunderstanding is that human life's being a basic good would entail that people are always interested in staying alive. But this claim has no such implication. The basic goodness of human life is compatible with emotional distaste for living; for emotion provides a source of motivation distinct from the human goods. Nor does the goodness of human life prevent other options, incompatible with life, from being choiceworthy. There are other goods besides life. Closely related is the fact that the goodness of human life does not establish any definite relationship between life and other basic goods. To say that human life is a basic good does not mean that life is any better than other basic goods. It means only that life is an irreducible ingredient in the ensemble of goods in which human persons find their full being.

One objection is that human life does not seem to function as a first principle of practical reasoning. If it were desired for its own sake, it would provide a sufficient reason for acting even when considered in abstraction from every other good. However, people desire not merely to live but to live well, not merely to survive in a vegetative state but to flourish self-consciously in a wide range of goods. Thus, deliberation seems to shape possible courses of action towards the promotion and protection of human life only as a necessary condition for the enjoyment of other goods, not as itself a basic human good.

This objection is plausible but depends on an equivocation. In one sense it is true, but in another sense false, that a basic good provides a reason for choosing *even when considered in abstraction* from other goods. It is true, by definition, in the sense that one need not look to any other good to find reason for desiring to serve or share in a good which is *basic*. But it is false if taken to mean that one can expect to find in any single good, however basic, all the richness and so all the rational ground for action present in the whole ensemble of human goods. This ensemble includes all the basic human goods, and so provides a rational ground for action which no single good, considered "in abstraction", that is, by itself, could possibly provide.

If one considers human life in abstraction (in the second sense) from other goods, for example, the vegetative existence of a person in irreversible coma, one is overwhelmed by the distance between this good and complete human fulfillment. Nobody wants to be in such a condition, and no decent person wants to see anyone else living like that.

But that fact does not show that life is merely instrumental. True, life is valued as a component of the integral human good. Whenever one can serve or share in the good of life by some particular action, the motivational power of that good is enhanced by the prospect of a life enriched by many other goods, all of them components of the fulfillment made up of the entire ensemble of goods. Still, this ensemble is not a supreme good beyond the basic goods. Only they provide reasons for acting. The ideal of integral flourishing only moderates the interplay of such reasons. Human life, like the other basic goods, provides its own reason for choosing actions by which one serves and shares in it.

No human good, considered apart from the entire human good, has the appeal that each of the components of that ideal enjoys when all of them are considered together. Who would find knowing the truth, or performing with excellence, or being a friend, or engaging in authentic worship, appealing by itself, apart from the others, in an existence (if that were possible) deprived of all the others? Little wonder, then, that life considered in abstraction from other goods is unappealing.

Still, even such an instantiation of the good of human life can by itself provide a reason for acting. Family members and health-care workers have chosen to give life-preserving care to persons they knew to be in irreversible coma. Not everyone would make such a choice or consider it correct. But the fact that some have made it gives evidence that life is a basic good – one which offers for choice an intelligible ground which need have no ulterior ground.

Another objection to the thesis that life is a basic human good is that it is not a good proper to the human person, but one which human beings share with other organisms. What is common to oneself and a carrot scarcely seems to be an irreducible component of one's full-being.

In reply it should be noted that life considered as common to all organisms is no more than an abstraction. In reality, life is as diverse as the variety of living things. All living things grow, but the growth of a carrot is a process altogether different from that of a dog. Not all living things see. Carrots live without sentience, but it is an important vital function of a dog.

To be able to do some of the things plants can do is not to be a plant; to be partly perfected by activities generically common to plants and animals is not to be partly a plant. Persons can do many of the things other animals and plants can do. But this does not mean that persons are brute animals or plants, nor that any part of the kinds of functions persons can perform belong to some brute animal or plant part of the human individual. Human beings are one species of organism. To belong to one species precludes belonging to any other, and any individual of a certain species is through and through of that kind.

The life of a human person, therefore, is a good proper to human persons. The growth of a human child is different from that of an individual of any other species. The sentience of a man differs from that of his pet. Brute animals live without deliberation and free choice, but these are among one's important vital functions.

It might seem that the preceding argument equivocates on "human life". In one sense, it includes all of the functions and activities of the person; in another sense, it is the minimal functioning without which there is no human organism. In the first sense, life may be a basic good in as much as it includes perfections proper to the person. But in the second sense, human life comprises only the vegetative functioning of the human organism.

The argument, however, does not involve this equivocation. For the word "life" is not used in two different senses, but in a single sense to refer to a reality which, like other basic human goods, can be instantiated more or less perfectly. When instantiated most perfectly, life includes vital functions such as speech, deliberation, and free choice, and is most obviously proper to the person. The life of a person in an irreversible coma is a very mutilated instantiation. But even so imperfect an instantiation of the good of life remains specifically human and proper to the person whose life it is.

VI. SERVING AS ONE'S OWN EXECUTIONER: SANCTITY OF LIFE OR DIVINE AUTHORIZATION TO KILL?

Even if the clarifications of the previous section establish that it is reasonable to think that life is a basic good of human beings, and therefore vindicate a key premise in my reformulation of the Thomistic argument against suicide, the question remains whether this good should be understood as demanding an absolute respect incompatible with any intentional suicide. In other words, it is necessary to consider premise (1) in my reformulation.

Donagan raises the difficulties with this premise in a pointed way. He notes a kind of suicide which seems not to be excluded by traditional arguments against suicide, including presumably Aquinas', although it is excluded by my reformation of Aquinas' argument. This is the case of a person who is publicly authorized to kill himself – namely, the case of a convicted capital criminal who is authorized or mandated to carry out the death sentence on himself. Such a person is not innocent in the relevant sense, and is authorized to kill a non-innocent person. So, this case would seem to be allowed by common morality, and Donagan suggests that Jewish and Christian refusal to allow it is to be explained religiously ([7], pp. 76–77).

This may appear to be a trifling challenge to the absolute prohibition of suicide, which could be met by a relatively insignificant adjustment of the argument so as to allow self-execution. For other, similar exceptions seem hard to conceive – cases, that is, in which the suicide is both authorized to kill and also in the process of violating just order, and so a proper object for his own authorized killing. The difficulty is in conceiving how the same person could both be violating just order and at the same time functioning as an authorized defender of that order. Such a person might have violated just order and expect to do so again, but authorization to kill without legal conviction of crime is not plausibly extended to past or future crimes.

Perhaps there are possibilities here, but they seem rare and may be fantastical. Within the framework of common morality the casuistry allowing a person to serve as his own executioner does not seem easy to extend. In particular, this casuistry would not justify one in intentionally killing oneself because one knew that one would commit a future crime, and even less would it justify self-execution on the ground that it would prevent one from causing future harms that were not violations of just order.

However, I believe that a consideration of this possible exception to the prohibition of suicide reveals significant tensions within Aquinas' position, and indeed more generally, within common morality. For, to the extent that my reformulation of Aquinas' argument is a plausible rendering of his thought, it would seem that neither the guilt or innocence of the person to be killed, nor the authorization of the killer, should make any difference to the justification of suicide. But, given Aquinas' general view of the ethics of killing, it is difficult to see how guilt or innocence does not, in some cases, make a difference. And if it does, it becomes difficult to see how Aquinas can maintain a consistent sanctity-of-life argument against suicide.

It is clear that Aquinas does not accept anything as strong as my reformulation of the first premise of his argument against suicide. Briefly, his position seems to be that no one, except on the basis of direct divine command, may intentionally kill an innocent person, but that on the basis of such a command one may intentionally kill an innocent person, and that persons having public authority may intentionally kill malefactors. I think, however, that the difficulty here is not that my reformulation of Aquinas' moral principle is inaccurate, but that his own position has difficulties that are scarcely kept from sight. His own argument against suicide, and his more general conviction that homicide is wrong not simply because of its injustice, surely require something close to the two premises in my reformulation. If I am correct in thinking that he would accept the proposition that human life is a basic good, then he must qualify the moral principle so that only innocent life is inviolable, and even that conditionally on God's not commanding one to kill an innocent. I am at a loss to see how Aquinas provides moral justification for the needed qualifications.

In response to the objection that killing is always wrong because inconsistent with the charity owed to every person, Aquinas held that the criminal loses his human dignity by his criminal activity. One's human dignity is explained as one's natural freedom and existence for one's own sake; when this is lost a person may be used for others like the brute animals. In fact, the sinner is worse than a brute animal (*ST* 2–2, 64, 2, *ad* 3). Aquinas' central argument for the justification of capital punishment is that an individual is related to society as part to whole, and so just as a part of the body can be excised for the sake of a whole, so can the criminal be killed for the sake of the common good (*ST* 2–2, 64, 2).

In discussing killing the innocent, Aquinas argues as follows: "Considering a human being in himself, it is licit to kill no one, because in any person, even a sinner, we should love the nature which God created and which killing corrupts" (*ST* 2–2, 64, 6; see also *ST* 2–2, 25, 6). He goes on to allow, however, that the killing of sinners is licit when considered, not in themselves, but in relation to the common good of society (*ST* 2–2, 64, 6). Another qualification is introduced: in considering the divine mandate to Abraham to kill Isaac, Aquinas allows that killing the innocent may be justified if there is a direct divine mandate to do it (*ibid.*, ad 1).

Aquinas also justifies suicide if done in response to a direct divine mandate (*ST* 2–2, 64, 5, *ad* 4).[17] In spite of his willingness to allow suicide (and killing the innocent) on the basis of a direct divine mandate, Aquinas is not willing to allow a person to serve as his own executioner. The reason is not a sanctity-of-life consideration, but the fact that one can have public authority to kill evildoers only because one can judge them and one cannot be a judge in one's own case (*ST* 2–2, 64, 5, *ad* 2).

There seem to me to be a number of difficulties with these claims, taken on their own terms. First, if we accept the obligation not to destroy human life when considered in itself, it is not clear why a negative relationship between the person whose life it is and the common good of political society has any tendency to remove the ground for not killing the person. This negative relationship does not remove or substantially change the person's nature which we are obliged to love in a way that excludes destroying it. Literally speaking, the criminal remains a human being, and punishment, properly speaking, presupposes the human dignity of the person punished.

Aquinas allows that a fuller description of an action can reveal morally relevant features of it which might require a change of its moral evaluation (*ST* 1–2, 18, 10). A fuller description can reveal that an action which, when described less fully, is judged morally indifferent is really either morally good or morally bad in virtue of features disclosed by that description; it can also reveal that actions judged impermissible when described up to a point are not really so, because features captured by the fuller description make clear that the putatively wrong-making features of the act do not obtain, or are not in the circumstances wrong-making features (*ST* 1–2, 94, 4). In the cases of killing, however, what makes the killing wrong in the less ample description includes the fact that a human life is involved, and the redescription of the action revealing the person's negative relation to the common good does nothing to remove whatever it is about human life that makes it wrong to destroy it. The cases of killing are not parallel to the example of returning borrowed goods, which is usually the right thing to do, but is not the right thing to do when the claimant of the goods means to use them for seditious purposes. In this example, what flaws the action is revealed as no longer obtaining when the circumstances are considered.

Second, the relationship between the individual and society is not a part/whole relationship which is easily assimilated to that between the human body and its parts. That suggests an instrumental view of the human person which surely cannot be sustained within common morality. Third, it seems too strong a condition for proper authorization to kill that one should be able to judge the evildoer. Executioners are usually not judges.[18]

Similar questions arise when one seeks to formulate the moral principle in Aquinas' argument against suicide. Perhaps it could be formulated as: no one should act with the intention of harming his own basic goods unless directly commanded to do so by God. This might be a conclusion from the norm that no one should intentionally harm the goods of innocent persons except when directly authorized by God, and the proposition that no one can be authorized to kill himself without direct authorization from God.

It seems to me that neither the norm nor the proposition, in the qualified form needed to allow what Aquinas is committed to allowing, has a plausible *moral* justification within common morality. The convictions underlying them are obviously religiously motivated, and I cannot see how they can be morally justified. For if they are understood as moral reasons, then God's authorization will not be regarded as arbitrary or necessarily incomprehensible. And if God's reasons are in principle comprehensible, whatever serves as God's ground for authorizing killing the innocent, or oneself if guilty, should plausibly extend beyond the cases he directly authorizes to other cases where his reasons can be seen to apply. Less radically, it is unclear why this authorization is not reasonably extended to those having political authority. For their concern is with the common good, which is better and more divine than any individual's good (see, e.g., *ST* 2–2, 141, 8). And, surely, such persons are capable of having knowledge of what is within their responsibility sufficiently like God's providential knowledge of human affairs generally to allow them, in some cases, to kill innocents, or themselves if guilty, for the common good.

Of course, such authorities do not, like God, have dominion over life and death, and are not reasonably regarded as executors of God's mandate (see *ST* 2–2, 64, 6, *ad* 1). However, if this consideration is decisive in rejecting the extension of the divine authorization to kill innocents or oneself, it demonstrates that the qualifications within the norm and the proposition are religious and not moral.

But Aquinas does suggest a moral reason for restricting the killing by public authorities to criminals. "The life of just people conserves and promotes the common good, because these people are the greater part of the population" (*ST* 2–2, 64, 6). His reasoning seems to be that public authorities can act within their authority when they act for the common good, and that when the lives of innocent people are considered in relation to the common good, there is no justification for killing them. So anyone who would kill the innocent without direct divine authorization would be acting outside his legitimate authority. One might wonder, however, whether Aquinas' argument is sufficient for him to sustain an absolute prohibition against killing the innocent when direct divine authorization is lacking. For there surely are cases where the premise in this argument does not hold – when just people do not comprise the greater part of the population. And further, unless refusing to kill the innocent is taken to be part of the common good, there are likely to be cases where those having responsibility for the common good might judge it necessary to kill some innocents.

It seems, therefore, that if we regard as religiously grounded Aquinas' exceptions to the prohibition of killing innocents based on direct, divine authorization, the sanctity-of-life elements in his arguments could easily be preserved with religious qualifications. But if we consider Aquinas' moral reason for limiting the authorization to kill innocents, these sanctity-of-life considerations seem forced out of the argument. For political authorities are prohibited from killing innocents not because they are human beings, but because the officials are outside their authority or are not acting for the common good. So even if the qualification in the norm underlying Aquinas' premise in the argument against suicide is religious, the limitation of that qualification is by a moral consideration, which seems incompatible with the sanctity-of-life strands in his arguments.

The other limitation on publicly authorized killing is expressed in the proposition that one may kill oneself only by direct divine authorization. The reason here also is a lack of authorization: one lacks in one's own case the knowledge to be a judge.

If this reason is as unpersuasive as I think it to be, it is hard to see how Aquinas can refuse to allow Donagan's exception. But perhaps Aquinas' reasoning on this point could be developed. Perhaps he could have argued that a person's special duties to self are such that it is necessary to avoid moral perplexity for that person to avoid public responsibilities that conflict with those duties.[19] However, since the common good is taken to be greater than the individual's private good, there seems to be no real perplexity here, at least when one serves as an executioner under orders and not as a judge.

Still, the limitation is based on moral reasons having to do not with sanctity-of-life but with conditions for public authorization to kill. These conditions form part of Aquinas' larger moral account of public authority, which includes a justification of publicly authorized killing. And this ethics of authorization appears to remove whatever function the sanctity-of-life elements in Aquinas' ethics might play in moral arguments about killing.

What perhaps conceals this fact is that Aquinas wishes to keep authorized killing limited to that done by public authorities; private persons may not intentionally kill even in self-defense. And he wishes to keep the killing by authorities strictly limited to what those having authority may do to certain restricted classes of persons, that is, criminals. This combination – that some killing is authorized, but strictly limited – is not based on sanctity-of-life concerns but may seem to approximate a sanctity-of-life position. What it requires is a special moral status for authorized killers, a status that will permit their authorized killing but provide no precedent for other killing.

My conclusion is that if there is a coherent moral view containing anything like a sanctity-of-life position and a justification of capital punishment and other publicly or divinely authorized intentional killing, then that view cannot be rooted in anything like Aquinas' version of common morality. If Aquinas' approach is to be one's starting point, then one must choose between the sanctity-of-life elements in his approach and those elements in his thought that allow authorized intentional killing.

It is, however, a mistake to think that in preferring those elements of Aquinas' view in which he allows for authorized killing, one is opting for the rational, moral component of Aquinas' view, and setting aside religiously motivated concerns about the sanctity of human life.[20] For Aquinas' convictions about such things as the reality of divinely authorized killing, capital punishment, and even the prerogatives of political society are all deeply influenced by his reading of the scriptures and his understanding of the religious tradition. And, although his concern for the respect for human life is frequently expressed in theological language, it is clear that this concern also has a foundation in his philosophical analysis of practical reason and human well-being. Thus, it is not at all clear that it is simply religious conviction to hold that a rational regard for the human person leads to an absolute prohibition of intentional killing.

Preserving the right of political society to execute convicted criminals and to kill intentionally those engaged in criminal activity is by no means a principle within common morality, nor is it anything like a datum which those who accept common morality must simply accept. Surely, it is no more of a given or a principle than the sanctity-of-life conviction it contradicts. Many who hold for some version of common morality reject any such preeminence on the part of political society, and in particular the subordination of the individual to political society on which Aquinas' discussion of authorized killing rests.[21] The traditional acceptance of exaggerated claims on its behalf can be explained in terms of the mistaken acceptance of the theocratic pretensions of dominant polities within societies influenced by common morality.

Still, one might wonder whether a different formulation of common morality might reduce the tension between sanctity-of-life concerns and the conviction that some intentional killing must be authorized. Donagan's discussion of suicide, from a Kantian point of view, suggests a negative answer. For it seems to me that Donagan comes close to excising sanctity-of-life considerations altogether from common morality's ethics of killing.

Donagan's discussion is important for a different reason. For Donagan, unlike Aquinas, but like most others within common morality now, does not ascribe to political society altogether unique prerogatives in the matter of killing. His discussion, therefore, reveals what happens when one allows authorized self-killing but rejects the Thomistic assumptions that keep authorized killings narrowly limited.

Donagan rejects Kant's absolute condemnation of suicide, since carrying out a lawful death sentence upon oneself is "not to dispose of oneself as a mere means to an arbitrary end" ([7], p. 77). He goes on to spell out other circumstances in which killing oneself is no more arbitrary than serving as one's executioner. He lists four kinds of circumstance: when suicide is necessary to avoid an enforced choice between denying one's fundamental practical allegiance or death or unendurable torture; when it is necessary to avoid a life unfitting for a rational creature; when it is done to ensure the lives or well-being of others; and when it is done to escape a life of natural degradation.

It seems that if self-execution is compatible with respecting rational nature, then suicide in the other circumstances Donagan outlines probably is, too. And these circumstances allow many kinds of suicide. For Donagan's view seems to exclude no suicide that is reasonably judged necessary to help others in a significant way. And, although Donagan plainly wants to limit significantly the self-regarding justifications, one wonders whether that limitation can be sustained.

In all three kinds of circumstance in which suicide is allowed because of a person's concern about his future prospects, there is a threat to something closely connected with rational nature: death or unendurable torture or denying one's most basic commitments, a life of human degradation, and a life of natural dehumanizaton. In all these cases one kills oneself to prevent what will compromise one's dignity as a rational creature.

Other cases come to mind which are similar to these in that one foresees a future threat to one's rational dignity. Why, for example, should it be a failure to respect rational nature if circumstances, including social convention, should make the alternative to suicide a life of social disgrace and humiliation? This need be neither a life of degradation unfitting a human creature, nor a life in a naturally dehumanized condition, nor is it a forced choice between denying one's practical allegiances or death or torture. But if suicide in these circumstances does not fail to respect human nature, why not also in this case?

Casuistry of this kind could, I think, extend the class of legitimate suicides in several directions from those which Donagan allows. If altruistic motives and prevention of compromises of dignity are grounds, why not other good reasons as well? The only clear limit is that suicide would be wrong when done at will, when one held one's life cheap. That, as I have noted above, is not a clear limit.

So, sanctity-of-life elements can be excised from common morality, and if the unique prerogatives of the state are also denied, one gets a view of the ethics of suicide that is much more permissive than common morality's greatest philosophical proponents would have allowed. Another alternative is to accept the sanctity-of-life elements within common morality and reject what is incompatible with them. That is a grim alternative for many within common morality, because it seems to put one at the disposal of bad people. But it is a moral option more appealing on closer inspection than is usually thought, and is not so obviously unreasonable as people often think.[22]

NOTES

[1] That is *Summa Theologiae,* Second Part of the Second Part, Question 64, article 3, sed contra. The work is abbreviated as "ST", followed by an indication of the part (1–2 means "first part of the second part"; 2–2 means "second part of the second part"), followed by a number indicating the question, followed by a number indicating the article, and, when necessary, followed by "*ad*" and a number to indicate a response to an objection, or "*sed contra*" to indicate the theological source for Aquinas' position. Thus the reference here would be *ST* 2–2, 64, 3, *sed contra.* Question 64 contains Aquinas' treatment of the ethics of killing.

[2] See *ST*, 1–2, 94, 2. Here Aquinas lists the basic goods of human nature that constitute the foundations of practical reason and the principles of morality. Among the goods proper to humans as rational is knowing the truth about God.
[3] Kant seems to have had a similar argument in mind: "God has forbidden it [suicide], because it is abominable in that it degrades man's inner worth below that of the animal creation" ([12], p. 153). See also the argument developed in the *Metaphysics of the Virtues,* cited by Donagan [7], p. 77.
[4] I do not maintain that Aquinas could consistently hold my reformulations of his premises. Indeed, I believe he could not. This fact, I think, reveals some of the central difficulties in the effort to develop a sanctity-of-life position from within common morality. See below, section VI.
[5] The only suggestions in this direction of which I am aware arise in discussions of the subordination of individual interests to those of the community, which Aquinas holds to have a more divine status than the individual; see, for example, *ST* 2–2, 64, 2, Aquinas' discussion of capital punishment. For a discussion of relevant texts, see [14].
[6] This is roughly how Donagan, extrapolating from Grisez's exegesis, understands Aquinas; see [7], pp. 63–65. For a discussion of the relation of the Kantian and Thomist formulations of the first moral principle, see [5]. See [11], pp. 183–189 for a formulation of the basic moral principle suggested by this reading of Aquinas, and pp. 216–222 for a derivation from it of the prohibition against acting contrary to instances of basic goods.
[7] There are other positions on killing in self-defense even within the Catholic tradition. Augustine held that a private person may not kill in self-defense; see *De Libero Arbitrio,* Book 1, Chapter 5. Aquinas' introduction of intentional considerations into the discussion may be seen as an attempt to save Augustine's position by limiting it to intentional killing in self-defense. Johannes DeLugo, S.J., an influential seventeenth-century moralist, rejected Aquinas' analysis and held that one may kill an assailant as a means to saving life; see *De Justitia et Jure, Disputatio X, Sectio VI,* numbers 148–149.
[8] See *ST* 1–2, 72, 1, for Aquinas' use of this distinction as the basis for distinguishing various aspects of any sinful act; for his application to specific cases see *ST* 2–2, 43, 3, and *ST* 3, 88, 4. For a substantiation of the interpretative claims in this and the following paragraphs, see [3].
[9] The quote is from *ST* 2–2, 64, 7, but the idea is also found several times in the general discussion of sin; see, for example, *ST* 1–2, 72, 8.
[10] In his account of the moral classification of human acts in *ST* 1–2, 18, there is no mention of the role of intention. There the emphasis is entirely upon the features of acts in virtue of which they either accord or do not accord with reason without consideration of how the agent's will bears on these features. I am uncertain whether these texts, and others in the same spirit, can be rendered consistent with the emphasis on intention in the discussion of sin and self-defense and in other contexts.
[11] The only context in which Aquinas seems to reject the relevance of intention for determining what is morally essential in a voluntary action having several morally relevant features is *De Malo* 2, 6, *ad* 6. The argument here is less than clear, but if it does reject the appeal to intention as a way to distinguish what is essential from what is not within a voluntary act, than it contradicts the analysis in *ST* 1–2, 78, 1.
[12] For a critique, based on Aquinas' discussion of killing in self-defense, of the efforts

of some proponents of the double-effect doctrine to draw the line between what is intended and what is *praeter intentionem* different from what I suggest here, see [2].

[13] Aquinas' analysis of willing is extensive and carried out in several works. The most developed analysis is in *ST*, 1–2, 6–17. I have summarized the most relevant parts in [3], pp. 650–654.

[14] See *In IV Sent., dist.* 4, 1, 1, *ad* 2: "quae praeter intentionem accidunt carent ordine ad finem".

[15] For a statement of the traditional doctrine of double effect, and further discussion on this point, see [4], pp. 528–532. It seems to me that most of Donagan's objections to double effect, as distinct from his defense of his own view of voluntary action, are based on this misunderstanding. See [7], pp. 122–127, 157–164.

[16] For a development of the application to killing of the prohibition against acts including the intent to harm human goods, and of the conditions under which death may be accepted as a side effect, see [8], pp. 297–319. For an analysis of the concept of murder and its relation to intentional killing of the innocent, see [1], pp. 18–21.

[17] Divine inspiration is said here to be necessary to justify Samson's suicide, and that of the holy women during times of persecution.

[18] For a development of these criticisms of Aquinas, see [10], pp. 66–73.

[19] See *ST* 1–2, 19, 10, for a suggestion in this direction.

[20] Here I am disagreeing with Donagan who, in so many words, holds that the sanctity-of-life elements in common morality's views about suicide are religious and the authorization of killing is rational. See [7], pp. 76–77.

[21] For a recent statement by adherents of common morality of a view rejecting the subordination of the individual to political society, see [15].

[22] See [8], pp. 297–319, for a statement of such a sanctity-of-life version of common morality, including an analysis of the extent to which the use of force defense of just order is justified without appeal to questionable views about the unique prerogatives of political society.

BIBLIOGRAPHY

[1] Anscombe, G.E.M.: 1982, 'Action, Intention and Double Effect', *Proceedings of The American Catholic Philosophical Association* 54, 12–25.

[2] Boyle, J.M.: 1977, 'Double Effect and a Certain Type of Embryotomy', *Irish Theological Quarterly* 44, 303–318.

[3] Boyle, J.M.: 1978, '*Praeter Intentionem*,' in Aquinas, *The Thomist* 42, 649–665.

[4] Boyle, J.M.: 1980, 'Toward Understanding the Principle of Double Effect', *Ethics* 90, 527–538.

[5] Boyle, J.M.: 1984, 'Aquinas, Kant and Donagan on Moral Principles', *The New Scholasticism* 58, 391–408.

[6] Boyle, J.M. Jr., and Sullivan, T.D.: 1977, 'The Diffusiveness of Intention Principle: A Counter Example', *Philosophical Studies* 31, 357–360.

[7] Donagan, A.: 1977, *The Theory of Morality*, University of Chicago Press, Chicago.

[8] Finnis, J., Boyle, J.M., Jr., and Grisez, G.: 1987, *Nuclear Deterrence, Morality and Realism*, Oxford University Press, Oxford.

[9] Grisez, G.: 1965, 'The First Principle of Practical Reason: A Commentary on The Summa Theologica, 1–2, Question 94, Article 2', *Natural Law Forum* 10, 168–201.
[10] Grisez, G.: 1970, 'Toward a Consistent Natural-Law Ethics of Killing', *The American Journal of Jurisprudence* 15, 66–73.
[11] Grisez, G.: 1983, *The Way of Lord Jesus: Volume 1: Christian Moral Principles,* Franciscan Herald Press, Chicago.
[12] Kant, I.: 1963, *Lectures on Ethics,* trans. by L. Infield, Harper and Row, New York.
[13] Kenny, A.: 1966, 'Intention and Purpose', *The Journal of Philosophy* 63, 642–651.
[14] Lee, P.: 1981, 'Permanence of The Ten Commandments: St. Thomas and his Modern Commentators', *Theological Studies* 42, 422–433.
[15] Vatican Council II: 1965, *Declaration on Religious Liberty (Dignitatis Humanae),* Vatican Polyglot Press, Rome.

St. Michael's College,
University of Toronto,
Toronto, Ontario, Canada

H. TRISTRAM ENGELHARDT, JR.

DEATH BY FREE CHOICE: MODERN VARIATIONS ON AN ANTIQUE THEME

I. INTRODUCTION

Few people choose death for its own sake. That is not to say that philosophers have not reflected on the meaning of life in a way that makes the possibility of suicide central to life's significance. One might think of the opening lines to Camus' *Myth of Sisyphus*, where he states, "There is but one truly serious philosophical problem, and that is suicide. Judging whether life is or is not worth living amounts to answering the fundamental question of philosophy" ([9], p. 3). Still, suicide usually plays a role within a larger context and is not seen as an end in itself.[1] This is underscored in classical accounts of suicide. One might think here of the description by Tacitus of the death of Seneca.

Then by one and the same stroke they sundered with the dagger the arteries of their arms. Seneca, as his aged frame attenuated by frugal diet allowed the blood to escape but slowly, severed also the veins of his legs and knees. Worn out by cruel anguish, afraid too that his sufferings might break his wife's spirit, and that, as he looked on her tortures he might himself sink into irresolution, he persuaded her to retire into another chamber. Even at the last moment his eloquence failed him not; he summoned his secretaries and dictated much to them which, as it has been published for all readers in his own words, I forbear to paraphrase ([33], pp. 391–2).

As a modern example one might take the death of Yukio Mishima.

Yukio Mishima the novelist chose to die a fanatic's death, and the most Japanese death imaginable. On November 25, 1970, accompanied by four cadets from his Shield Society, he paid a visit to the commandant of the Japan Self-Defense Force. On his signal, the cadets seized the commandant and held him at swordpoint, while Mishima demanded through the barricaded office door that the 32 Regiment he assembled in the courtyard to attend a speech. At a few minutes past noon, he stepped out to the balcony and exhorted the soldiers to rise up with him against a postwar democracy that had deprived Japan of her army and her soul. He had intended speaking for thirty minutes, but since his words were inaudible above the jeers and hisses of the eight hundred angry men, he stopped after just seven. Then he withdrew to the commandant's office and committed *seppuku* (hara-kiri). When he had driven the blade into his left side and drawn it across his abdomen, he grunted a signal to the cadet standing behind him; the cadet beheaded him with a long sword, completing the ritual.

Baruch A. Brody (ed.), Suicide and Euthanasia, pp. 251–280.
© 1989 *Kluwer Academic Publishers*.

This scene is dramatized in the movie *Mishima*, directed by Paul Schroder ([25], p. ix). Both Seneca and Mishima gave meaning to their lives through their deaths. Their way of exiting from this world affirmed the values for which they lived and gave a legacy to those who valued them and their commitments.

In many respects, choosing one's way of dying is like writing a deathbed speech. One speaks through such choices to others, affirms certain values, and sets into motion trains of causes that continue to influence the world and those in it after one is gone. The choice of one's death can thus be a matter of significant moral importance and can be an affirmation of meaning, not of despair. This is not to deny that most who commit suicide do not do so as a free choice, but rather as a behavior evoked by mental illness or disorder. For this vast majority, treatment, support and care are appropriate. Quite often this should include involuntary commitment. Where free choice no longer exists and individuals are overborne by mental illness, there is not autonomy to respect, but care to be given.[2] Still, there are circumstances under which rational individuals will choose to die at their own hands or with the help of others. It is that minority of cases, involving rational choice,[3] which is important for understanding the significance of suicide, for in those cases it takes on a moral significance because of the choices of the agent.

The last two decades have seen a remarkable interest in rational suicide and euthanasia. Numerous books and articles have been published and various societies organized to support legal reform. The latter have ranged from lobbying for the enactment of natural death acts to frank support of laws that would allow voluntary euthanasia. Even do-it-yourself guides have been written. If one seeks sources for all of this interest, a number can be advanced. First, our societies are now much less in the grip of the traditional Western religious orthodoxies, which condemned suicide. Secular reappraisals of the rights of individuals to control their own lives, including the circumstances of their own deaths, have introduced discussions of voluntary euthanasia and suicide as a part of an extensive list of rights to control oneself and to engage in consensual acts with willing others. The rationale for forbidding rational suicide and voluntary euthanasia has thus come under critical reexamination as a part of a general appraisal of the scope of state authority.

This growing recognition of individual rights and limited state authority has occurred just as medicine has succeeded, either directly or indirectly, in prolonging the lives of individuals. More people are now living longer in both absolute and relative numbers than at any time in the past. Significant numbers of individuals will survive into their 80's and beyond. Life is longer for more people than ever before. However, such success is often a mixed blessing. Individuals now survive castastrophic accidents, diseases, and other mal-events, only to be faced with severe circumscriptions of their abilities, both physical and mental. Human bodies can be kept alive for years in the absence of whole-brain death and in the absence of consciousness.[4] The process of dying can be extended weeks and months. Diseases that would have more quickly killed in the past now can leave individuals totally paralyzed but fully alert, as in the case of amyelotropic lateral sclerosis. Though many individuals can look forward to surviving into their 80's, an appreciable percentage of them will do so while suffering severe forms of senile dementia, such as Alzheimer's disease. Though medicine may succeed in dramatically extending life, it does so at times without having secured a quality of life acceptable to many who will have to live it.

This second theme concerns the control of technology. If a medical intervention is likely to prolong one's life but runs a substantial risk of doing so at an unaccepted quality of life, one may be afraid to take the risk of being saved if one is exposed to the danger of being locked into an unacceptable quality of life, from which one can find no easy exit.[5] There are cases when individuals may wish to avoid a life-saving procedure because they fear that they will not be allowed to turn down the life saved, should the quality not be acceptable. In such circumstances, the proscription of suicide, assisted suicide, and voluntary euthanasia has the dual costs of (1) discouraging some from a reasonable attempt to achieve a life of quality, while (2) condemning to protracted suffering those who find the life saved unacceptable.

The decision to accept treatment may secure only a very short survival burdened by significant suffering. Patients or their families may decide in favor of intubation (artificial support of breathing) rather than allow the agony of gasping for breath as respiratory failure progresses. As a result, life is prolonged, often with poor quality of life, if not actual pain. Some would undoubtedly decide against intubation in the face of certain death and in favor of a merciful dosage of an opiate to accelerate death or directly accomplish it. Such choices are not usually offered to patients because of the fear they would involve criminally proscribed acts of assisted suicide or voluntary euthanasia. Such hesitations exist, though in many circumstances the provision of sufficient medication to control pain has been acceptable to moral theologians, and perhaps fairly generally, even if it may increase the risk of dying sooner, as long as death is not directly intended, and the pain relief does not result from a dose sufficient to kill rather than simply to blunt pain.[6]

Finally, the interest in suicide, assisted suicide, and voluntary euthanasia has been spurred by the costs of protracted survival, especially when it is of no use to the person surviving. There is evidence to indicate that 20% or more of the patients who use the $15 billion invested in intensive-care units (ICUs) are receiving such treatment, though there is no likelihood of survival [2]. Some of the funds could be saved simply by discontinuing treatment when it offers no benefits and only prolongs dying and suffering. There is a difficulty, however, in that individuals do not die as in Verdi operas, after singing their last aria. Even under the best of circumstances there are usually hours if not days of deterioration of function and often agony and misery before death occurs. However, were assisted suicide and voluntary euthanasia legally available, one could choose an easy exit when death is imminent. One might not be attracted to particular medical interventions that promise the relief of only some suffering but no cure and thus piecemeal bring one to extending one's life, often in a setting tantamount to a high-intensity-care hospice.

Problems with medical costs are not restricted to high-intensity-care settings. The finances of retired couples and the inheritances of families are often invested in long-term nursing care, which offers little, if any, contribution to the quality of life of the person receiving the care. One might think here of individuals with very severe senile dementia, as in the last stages of Alzheimer's disease. Such care is often not what would have been wanted by the person to be sustained. Still, the prospect of a slow, albeit relatively painless, death after discontinuing hydration and feeding may seem to many an unacceptable exit for themselves, or for their loved ones, even if the loved ones left instructions to stop all support. To many, a cleaner, quicker exit may appear both more dignified and more humane.

Medicine has thus returned us to classic themes about suicide. Our capacity to prolong life has given us control over our dying, to the point that we must now often choose when we are to die. But often simply stopping treatment will not provide the best death, or even perhaps an acceptable death. Though these questions are often raised in highly technological settings, they have ancient roots and are part of themes that recur in philosophy and literature. One might think of Seneca's observation that when "one death involves torture and the other is simple and easy, why not reach for the easier way?… Must I wait for the pangs of disease… when I can stride through the midst of torment and shake my adversaries off?" ([32], pp. 204–205). Similarly, Friedrich Nietzsche has Zarathustra remark concerning voluntary death that

Many die too late, and some die too early. Yet strange soundeth the precept: 'Die at the right time!' Die at the right time: so teacheth Zarathustra ([26], 1, 21, p. 75).

We are brought then to a reassessment of the moral assumptions of the public policies that support criminalization of suicide, assisted suicide, and voluntary euthanasia.

II. THE STATE MONOPOLY OVER LIFE AND DEATH

Western societies have made strong claims concerning their rights over their citizens. They have regulated the production of children through regulating access to contraception, sterilization, and abortion. So, too, the circumstances of death have been under strict state control. Western governments have generally forbidden individuals to kill themselves or to kill competent individuals who request death. These state intrusions spring from a view of the state and of the individual's relationship to society that is bolstered by a particular understanding of morality and of the relationship between morality and the use of coercive state force.

To appreciate these claims of state authority over individual choices of death, consider how the authority of the states in general has been understood. For the West, two cities have had metaphorical significance: Jerusalem and Athens. Jerusalem can be used as a metaphor for a theocratic understanding of morality and public policy. The Deity reveals His truth, appoints His agents, and they enforce divine rules with divine authority. This understanding of the sources of state authority is embraced, for example, by Saint Paul: "the powers that be are ordained of God" (Romans 13:1). *Pace* Mohammed, Teheran in this classification is an instance of Jerusalem. In contrast, Athens can be used as the metaphor for a rationalist understanding of morality and public policy. Reason is seen to be sufficient to determine a concrete vision of the good life, and those who discover it are then authorized to impose it by force. Their use of force can be justified as rational, and those who resist condemned as irrational. Paris of the Revolution and Moscow of the 20th century are instances of Athens.[7]

These two cities offer two modes for resolving controversies regarding morals and public policy. The first appeals to a special revelation to authorize force, and the second to reason to give a conclusion to a controversy through sound argument. However, the more one is skeptical regarding the existence of such divinely provided guidance or regarding the limits of reason, the less plausible either solution will seem. One is confronted with a question in controversy theory. The issue is one of bringing controversies to closure by appeals to other than brute force. Insofar as individuals do not share in the consensus of a common religious belief, including the divine roots of state authority, appeals to religious consideration will appear to those without faith or with a different faith as an appeal simply to force in order to support private interests. There are other gods. Besides Jerusalem, there are Benares and Kyoto. Appeals to particular inspirations cannot form the basis for the peaceable resolution of disputes in pluralist societies. The failure of Jerusalem as a basis for public policy can be found in the Reformation, the British Civil War, and the Thirty Years' War on the Continent. After the Pax Westphalica and the Great, Glorious, and Bloodless Revolution, people can be said, with a number of qualifications, to have agreed to disagree with respect to matters of divine inspiration. This is not to say that remnants of religious justifications for the proscription of suicide, assisted suicide, and voluntary euthanasia do not remain. They are present as traces of a once dominant orthodoxy. One might think here of William Blackstone's condemnation of suicide as "invading the prerogative of the Almighty, and rushing into his immediate presence uncalled for" ([3], IV, p. 189).

The hope was to replace the failed city of faith with the city of reason. The Enlightenment sought a rational grounding for a concrete view of the good life and for public authority. Among its chief exponents was Immanuel Kant, who endeavored to establish the fabric of the moral life and warrants for state coercion on the basis of rational arguments. On the basis of what he took to be sound rational arguments, Immanuel Kant condemned suicide and developed three sorts of arguments against it. In some passages he contends that suicide is immoral, for if one destroys one's life out of self-love (e.g., from a wish to avoid pain), one wills a contradiction in wishing to destroy the self one loves. "Now we see at once that a system of nature of which it should be a law to destroy life by means of the very feeling whose special nature it is to impel to the improvement of life would contradict itself, and therefore could not exist as a system of nature; hence that maxim cannot possibly exist as a universal law of nature, and consequently would be wholly inconsistent with the supreme principle of all duty" ([22], pp. 39–40, *Ak* IV, p. 422). In this passage, Kant appears to regard the contradiction as a formal one. In other passages, it is suggested that if suicide were universalized, this would lead to the de facto obliteration of morality itself. "To destroy the subject of morality in his own person is tantamount to obliterating from the world, as far as he can, the very existence of morality itself" ([23], pp. 83–84, *Ak* VI, pp. 423–24). In yet other passages he condemns suicide because it involves using oneself as a means. "He who contemplates suicide should ask himself whether his action can be consistent with the idea of humanity *as an end in itself*. If he destroys himself in order to escape from painful circumstances, he uses a person merely as *a mean* to maintain a tolerable condition up to the end of life" ([22], p. 47; *Ak* IV, p. 429).

There are responses to each of these arguments. First, to commit suicide is not necessarily to will to destroy oneself. In choosing suicide or euthanasia, one wills to have one's self exist through a smaller set of states of pain than through an alternative large set, which would include a painful death. One need not intend to destroy oneself, but only to preclude certain future states of pain for oneself. For example, the person committing suicide or asking for euthanasia need not reject out of hand future pleasant states in an after-life.

As to the issue of undermining the very possibility of morality, one must distinguish between acting against the very concept of morality and peaceably precluding the existence of a moral community. To use unconsented force wilfully against the innocent is to reject the very notion of a moral community. Even if all decide to not reproduce or to commit suicide simultaneously on cue, they are not rejecting the very notion of a moral community, and they may not even be intending that no such community exist. They may be perfectly at peace if moral beings evolve elsewhere; they simply do not wish themselves or their offspring to exist who would then constitute an instance of a moral community. Also, to affirm suicide in certain kinds of circumstances is not to affirm it in all kinds of circumstances. One must be clear as to which affirmation of suicide is available to be universalized.

Finally, though suicide or voluntary euthanasia may involve using one's self or others as means, they are not instances of using oneself or others as means merely, as long as there is consent. It is interesting to note that Kant regards even selling one's hair as an instance of immorally using oneself as a means (*Metaphysical Principles of Virtue, Ak* VI, p. 423). He uses the same argument to condemn masturbation (*ibid.*, p. 424f.). For his arguments to have succeeded, he needed to establish not only duties to other individuals, but duties of individuals to themselves. This seemed feasible to Kant because he confused freedom as a value with freedom as a side-constraint. But freedom as a value is not a necessary condition for the possibility of a moral community based on mutual respect. And if all that is required is the respect of freedom, not the valuing of freedom, then free consent renders the taking of a life morally licit.[8] Consequently, *ceteris paribus*, rational suicide and voluntary euthanasia, not to mention duelling, may not be forbidden on the basis of general moral arguments such as Kant's. Kant, as others who have attempted to derive a concrete view of the moral life from rational arguments, needed to smuggle into his notion of a rational agent a particular set of values. The Enlightenment hope collapsed in this respect because it was asked to do more than it could. Though it could deliver the minimal conditions for mutual respect, it could not provide a justification for a particular view of the good life. It could show that promises must be kept, though it could not show which promises ought to be made. It could show that people may not be shot for sport without their consent, but it could not show whether a consensual mutual hunt unto the death is morally opprobrious.

Even if both Jerusalem and Athens fail, there is still Reykjavik, or, if you like, Washington-on-the-Brazos. A significant element of English common law can be traced back to the notion that the law is not derived from either God or reason, but from an agreement by men to a minimal set of conditions required for a community that allows for mutual respect. An example of such a community can be found in elements of the old pagan Icelandic community. With the establishment of the Althing at Thingvellir in 930, a legal structure came into existence that relied primarily on mutual agreement. Basic rights were those prerogatives that individuals had not given over to the general community. Basic rights were not regarded as given by God or by reason, but by the limits of communal authority. A similar set of assumptions can be seen at work at Washington-on-the-Brazos in the establishment of the Republic of Texas. No significant appeal was made to inalienable rights given by God, but to the limits of communal authority [14]. As a consequence, in the early days of the Republic, not only were suicide and aiding and abetting suicide not criminalized, but neither was duelling.[9] Within such a context, the gravamen, the evil of murder, was not the taking of someone's life, but the taking of his life without his permission. *Volenti non fit injuria.*

I have introduced the metaphor of the three cities to underscore the different ways in which communities have approached the idea of state authority. Below I will look in greater detail at the reasons why Reykjavik is the most extensive state that can be morally justified.[10] Here in anticipation it is worth noting that the arguments for rights of individuals to control their deaths are best derived in a negative fashion. This is so because it is difficult to establish a particular view of the good life and/or the moral authority to impose it by force. As a consequence, individuals possess generally defensible rights of self-determination. The more such rights exist, the more the moral life will need to be articulated within two spheres or dimensions. The first will involve those restraints that are justifiable in terms of sound rational arguments across competing visions of the good life. These are likely to be few and will be of the sort: "one should not break promises once made" and "one should not kill people without their permission". The second sphere will involve appeals to particular visions of the good life, which will not be rationally justifiable in a final or decisive fashion. Instead, there will be common understandings of the good life, which certain individuals will find plausible. It is through such particular visions or understandings of the good life embraced by religious grace and with religious conviction that people determine what promises to make, when it is worth living, and when, if ever, one may or should take one's own life. It is because of this complexity that it will often be possible to argue that people have a right to do X, though it is wrong. One may very well be able to establish that individuals have a right to commit suicide, assist suicide, or to engage in the provision of voluntary euthanasia, in the sense that states have no moral authority to use force in preventing such actions. Still, one may hold that there are suggestive or religiously grounded but not rationally definitive moral arguments to hold that suicide, assisted suicide, or voluntary euthanasia is wrong. One may want to witness on behalf of one's particular view of the good life and good death, though one must be tolerant of those who will not convert.

To understand this predicament, one must reflect on what is necessary to bring closure to a moral argument.[11] Disputes regarding concrete visions of the good life turn on particular rankings of values and harms. To resolve disputes regarding whether liberty is of greater value than status or wealth, one will need to appeal to a canonical ordering of values. However, such disputes exist because people defend incompatible rankings. One will not be able to solve such disputes by asking what an impartial or disinterested observer would choose. If the disinterested observer or group of disinterested observers does not have a particular moral sense in terms of which to rank benefits and harms, the observer will then be so disinterested as to be useless. However, as soon as the observer becomes useful, the observer becomes too particular. One will have attributed to the disinterested observer a particular moral sense, a commitment to a particular hierarchy of values. Nor will appeals to consequences help in resolving moral disputes. One will need to determine how to weight the consequences and compare them, which itself presupposes a particular hierarchy ranking harms and benefits. Nor will appeals to particular intuitions help, for any intuition can be met by a contrary intuition. The result of such reflections is that one becomes quite skeptical concerning the possibility of discovering an answer to the question of the concrete character of the good life.

If one cannot discover the concrete character of the good life, then one cannot discover when and under what circumstances it is proper coercively to forbid rational suicide or aiding and abetting suicide. Against those who would hold that life has an overriding value or importance, others will respond that their disinterested observer or moral intuitions support the proposition that liberty has an overriding value defeating the claims of the value of life. The result of these reflections is not nihilism. If one seeks the necessary condition for the possibility of a practice of blaming and praising, for the very possibility of ethics as an alternative to resolving disputes by force, mutual respect will be sufficient. One will still be able to negotiate a common view of the good life, insofar as the participants are willing. The results will have authority from the process of mutual agreement, even if an appeal to either reason or God will not in principle be definitive in a secular pluralist society. In addition, there will be a justification for condemning those who use force or who violate contracts, for they will have acted in a way that rejects the very possibility of mutual respect and therefore they cannot protest when they and their wishes are not respected.

This can be stated a different way. If justified authority for establishing a concrete view of the good life cannot be derived from an appeal to private intuitions (e.g., divine revelation) or from sound rational arguments, then there is still a straightforward way in which authority is derivable, namely, from free negotiation and consent. Respect of freedom functions here not as a value but as the source for a general secular ethics and as the basis for a moral community composed of moral strangers (i.e., individuals with different concrete views of the good life and death). One is saved from nihilism at the price of becoming a libertarian with regard to public policy in a secular pluralist society. One can always presume authority to engage in those actions necessary to protect individuals against unconsented actions. Rape, murder, burglary, and breach of contract can be punished. However, free love, duelling, gifts of money, and contracts under which one promises to kill oneself on cue from the Director of the C.I.A. cannot be condemned, according to the canons of a general secular peaceable morality.

The limits of human reason thus lead to dramatic limitations on the plausible authority of states, communities, and professions – and consequently to a new grounding for individual rights. Insofar as one wishes to speak with justification of blame and praise and to have a mode for resolving controversies without appeal to force, one will need to respect the innocent in the sense of asking for their cooperation, rather than coercing their service. Again, to use force against the unconsenting innocent would be to reject the moral community and to make oneself an enemy of peaceable persons generally. Insofar as it will not be possible to discover a canonical moral view, one will need to gain the agreement of others in fashioning common endeavors. However, there is no need for individuals to negotiate or to agree to particular contracts or relationships. Individuals are free to act wherever societal or communal restrictions have not been created, by explicit agreement, with the exception of those limitations that are integral to mutual respect, to the minimal conditions for the possibility of a moral community.

Those who use force to resolve a dispute will bear the burden of proving their authority. On the one hand, it is they who use force and therefore raise the question of authority. On the other hand, the minimal notion of a moral community requires mutual agreement, unless there are special grounds to make individual explicit consent unnecessary (e.g., a robber implicitly rejects the notion of mutual respect and therefore cannot protest when defensive or punitive force is used). Force users, whether individuals or states, claiming to have the authority to impose a particular view of the good life on non-consenting others, as a consequence of the minimal conditions for mutual respect, will need to be able to demonstrate that (1) their view of the good life is *the* canonical view of the good life, and (2) they possess authority to impose it on others without their consent. A final condition of prudence must be considered as well. (3) Even if one came to know the canonical view of the good life and had the moral authority to impose it on unconsenting innocents, it would not be prudent to impose it, were the costs of the imposition greater than the benefits to be derived.

The more one is skeptical regarding the capacity to discover *the* canonical view of the good life, and/or the authority to impose it by force, the more individuals have the right to do with themselves as they wish. Basic rights reflect the plausible limits of the authority of others. To talk of moral rights to commit suicide, to ask others to aid in one's suicide, to aid others in their suicide, to ask others to kill one under certain circumstances, or to respond to such requests is to question the plausible limits of the moral authority of third parties to intervene. It is important to underscore once again that just because people have a right to do X, it does not follow that it is good to do X. Rights in the sense just outlined indicate the limits of the authority of others to use innocent individuals without consent. Such rights are deontological.[12] They set limits on the actions of others independently of whether such limits, according to most concrete understandings of the good life, are seen to be beneficial or harmful to individuals or to society as a whole. Such rights set constraints on the extent to which people can be drafted without their permission in the service of others.

As a result of the limits of moral reasoning and the consequent limits of societal authority, people are at liberty to do many things or to omit many actions, even if those choices are held by many to be morally wrong, unwise or imprudent. If people are not owned by others, they are at liberty to injure themselves and consenting others, and to fail to come to the assistance of those in need. They may engage in actions that many hold will damn their souls. To be free is to be at liberty to pursue one's own view of the good life, however deviant, with consenting others in the absence of conclusive arguments to establish the basis for restraint. In this light consider the following examples of free choice: an individual who has great talents and is likely to become an excellent concert pianist decides instead to join the French Foreign Legion; a man divorces his wife after the birth of their fourth child, reminding his wife that, when they got married, he had warned her that he was not sure whether marriage was forever (let us also presume that he was a plausibly successful husband and father and that his departure is a loss to the wife and children); a 16-year-old girl decides to have as many children out of wedlock as possible, hoping for the support of third parties. In each of these cases, the individuals may have the right to engage in these actions, though they involve the loss of important goods.

In this paper, I will satisfy myself with exploring the limits of plausible state authority to restrict individuals who may wish to commit suicide, to assist suicide, or to provide voluntary euthanasia. My attempt will be to show what rights people have in these matters over against the state and society. I will not be trying to establish whether such actions are morally right or wrong. My particular vision of the good life and death is not a part of this essay.

III. THE RATIONALES FOR PROSCRIPTION

Anglo-American law has traditionally proscribed suicide [11], [18]. The punishment levied against the suicide included under English law forfeiture of all land and burial in a public highway. Though such sanctions were never adopted in the United States [8], suicide was considered at common law to be a crime and an act of immorality.[13] The proscription of suicide has been justified by appeals to various moral considerations and state interests.

1. The state as protector of the Deity's interests

Blackstone ([3], IV, p. 189) advances offense to the Deity as his first justification for the proscription of suicide. If the king reigns by divine right, or if the government is in place by Divine authority, it is not implausible that the king or the government has authority to prevent immoralities that injure God's interests. One must recall here as well the long tradition of felony prosecutions in the United States for so-called crimes against nature that has similar origins.[14] But insofar as one lives in a secular-pluralist society, appeals to the interests of the Deity will be inappropriate, for there will not be a commonly shared view regarding the existence of God, much less his interests. Such appeals will have no general authority that can be justified to men and women who are moral strangers. In this regard, justification of state coercion through appeals to the interest of the Deity will be indistinguishable from appeals to mere force.

2. Citizen or subject as property of the nation or king

William Blackstone lists as a second reason for forbidding suicide the king's "*interest in the preservation of all his subjects*" ([3], IV, 189). This understanding of the relationship between the citizen and society presupposes what Robert Nozick characterizes in criticism as ownership of the people, by the people, and for the people ([27], p. 290). Such a relation requires a very powerful justification. The arguments in this paper indicate that arguments on behalf of state ownership will be very difficult, if not impossible, to secure. Nor will appeals to some past contracts bind current citizens, unless one has a very robust theory of hereditary slavery, such that agreement on the part of one's ancestors to a particular constitution or government entails one's agreement as well. Appeals to majority votes as procedures for resolving such disputes will also be inappropriate, unless one has actually agreed to abide by such results. There is no more a divine right of majorities than there is a divine right of kings. In short, absent some actual agreement through which persons convey to the state rights in their bodies (e.g., as occurs when one joins the armed forces), claims that the state may forbid rational suicide, aiding and abetting suicide, or voluntary euthanasia will not be able to be secured.

3. O mores! Are there good ones the state may impose?

Similar considerations will bear against claims that the state has the right to forbid suicide, aiding and abetting suicide, or euthanasia in that they affront the moral values of the community. Such claims require an appeal to a particular moral sense, which is likely to be endorsed only by particular communities, not by all citizens, as a part of their participation in the societal structure. One might think here of analogous arguments that have been made in other areas of consensual conduct. Consider, for instance, the holding of the New York Appeals Court regarding the right of competent, consenting adults to engage in deviant sexual acts in private.

> We express no view as to any theological, moral or psychological evaluation of consensual sodomy. These are aspects of the issue on which informed, competent authorities and individuals may and do differ. Contrary to the view expressed by the dissent, although on occasion it does serve such ends, it is not the function of the Penal Law in our governmental policy to provide either a medium for the articulation or the apparatus for the intended enforcement of moral or theological values. Thus, it has been deemed irrelevant by the United States Supreme Court that the purchase and use of contraceptives by unmarried persons would arouse moral indignation among broad segments of our community or that the viewing of pornographic materials even within the privacy of one's home would not evoke general approbation. ... The community and its members are entirely free to employ theological teaching, moral suasion, parental advice, psychological and psychiatric counseling and other noncoercive means to condemn the practice of consensual sodomy. The narrow question before us is whether the Federal Constitution permits the use of the criminal law for that purpose [28].

The Court was drawing an implicit distinction between those actions the state may undertake in securing the peaceable community and those actions tantamount to imposing a particular moral viewpoint. This distinction rests on the difference between that tier of morality that can be justified to all moral agents as such, and the tier created through the particular actions and beliefs of individuals.

One should note that views regarding the importance of the sanctity of life are integral to particular views of the good life and depend on the endorsement of a particular moral sense. Against any claims regarding the importance of the sanctity of life, counterclaims can be advanced regarding the sanctity of free choice. Another way of putting this is that killing another person cannot be shown to be a *malum in se*, at least in terms of general philosophical arguments that do not already presuppose a particular ideological or religious viewpoint. What is wrong with murder is taking another person's life without permission. Consent cures. The competent suicide consents.[15]

4. The preservation of life

Allied with concerns for the sanctity of life are assertions that the state has an interest in the preservation of life. Insofar as the preservation of this interest involves the use of state force to support a value attributed to the sanctity of life, it is without moral authority, as the foregoing arguments indicate. Such a policy requires the imposition of a particular view of the good life by force. Insofar as the preservation of this interest involves a view that individuals are resources for a society or government, it requires a view of communal property rights in individuals. As was discussed above, this view presupposes arguments that do not appear to be sustainable, in that substantial societal authority must be secured for such interventions. Indeed, the reflections in section II undermine the notion that states or societies have a right to use force to compel innocent individuals to maintain the state's or society's existence.

5. The slippery slope and the police role of the state

Even if one grants that the state may not forbid suicide, attempted suicide, or euthanasia, one can still show that the state may come to the defense of those who are being killed without their valid consent. Any individual may render such aid, and so, too, may the state. The state may therefore validly develop means to protect individuals from killing themselves not through their free choice, but because of disease, or to protect individuals from involuntary euthanasia. These considerations can lead to at least weak paternalistic arguments, under which the state or others claim a right to interfere in order to determine whether a would-be suicide or recipient of voluntary euthanasia is choosing freely.

The more interesting question is whether the state may preclude rational suicide, aiding and abetting suicide, or voluntary euthanasia on the grounds that such constitutes a slippery slope, which will slide us down to a society where involuntary euthanasia and other forms of murder will be more frequent. There are two sorts of rebuttals to such concerns. The first is to deny the empirical claim made that such license will lead to a state of affairs similar to, if not worse than, Nazi Germany. The second involves a reply in principle that, even if such is the case, as long as one did not directly cause the evil outcome, others should determine how to defend themselves against such untoward consequences without forbidding consensual acts.

The first reply has a very strong factural basis, given the case example of Texas, which never forbade suicide and up until 1973 did not forbid aiding and abetting suicide. There is no evidence that such a legal vacuum led to an increased frequency of suicide or of murder [11]. There are in the end relatively few people who actually wish to avail themselves of rational suicide or to have their suicide aided and abetted. The rational actions of those who do, appear to have little, if any, adverse impact. In particular, the comparison with Nazi Germany is specious. The difficulties with Nazi Germany lay not with allowing voluntary euthanasia, but with seeing the state as having the authority to determine when men and women should live and die. As such, Nazi Germany labored in the shadow of traditional appeals to authority from the Herr Gott and from reason. The evils of Nazi Germany derived from the view that individuals only had those rights over themselves granted to them by the state. In contrast, Washington-on-the-Brazos (i.e., Texas) developed in the light of the pagan traditions of Reykjavik, where the state only had that authority conferred on it by individuals. States on their own have no rights to confer to anyone. Rational suicide, assisted suicide, and duelling thus gave a moral instruction directly contrary to that of Nazi Germany. Indeed, if one worried about slippery slopes, perhaps one should worry about policies that would forbid voluntary euthanasia and rational suicide because of a concern for the sanctity of life. Those well-meaning proscriptions may in the end lead to governments like Nazi Germany in that such policies presume that the state has the power to impose by force particular views of the good life and good death on the unconsenting innocent. One should recall that Nazi Germany did not allow but rather forbade abortion on request.

But if one did have data to show that the free actions of individuals would cause others to engage in immoral actions, is one obliged to forego one's morally permissible actions? What if one concludes that printing, distributing, and reading the Bible inevitably lead to the development of groups of fanatics who impose by force particular understandings of the religious life? What if one agrees that in the past such impositions have been among the most brutal, if not the most brutal, known to man? Is it immoral to print, distribute, and read the Bible, even if one does so for the purpose of making it available to peaceable individuals with tolerant religious sentiments? Must one imagine working out a system of registering Bibles as some would register guns, and only allow those access with a low fanaticism score on specially devised psychological tests? If such precautions will not be sufficient, must one categorically forbid the printing and reading of Bibles? Similar concerns underlie the denouncements by totalitarian governments of those who engage in intellectual thought pollution. May one pollute the intellectual world as one does the physical world? Do the innocent have a right to protect themselves from intellectual pollution, as they do from environmental pollution?

If the printing and distribution of Bibles are done with the malevolent intent of creating a fanatical movement to suppress the rights of unconsenting individuals, one may have a strong argument for intervening. However, there are countervailing considerations, chief among them that the creation of a latter-day secular Roman and Universal Inquisition, along with an Index Librorum Prohibitorum, would itself run the risk of circumscribing legitimate personal liberties and therefore might be as bad as, if not worse than, the evils against which it was designed to protect. But this claim on behalf of a robust right to free speech and free press is still an empirical one, which may in certain circumstances not hold. What if one has evidence to show in certain circumstances it is likely to be false? In limited circumstances the recognition of such grounds for restricting liberties appears to cause little difficulty. One might think here of the argument of Justice Oliver Wendell Holmes in 1917 in *Schenck v. United States*, where Holmes held that one could circumscribe free speech only when there was a "clear and present danger" that substantive evils would be caused, which the government has a right to prevent [31]. This may be interpreted as an instance of the right to stop individuals from malevolently or negligently causing dangerous behavior, not simply suggesting an evil action. But what of the non-malevolent dispersion of ideas, which may give individuals a reason to act in a particular way, not simply cause them to behave dangerously, as shouting "Fire!" might in a crowded theater. In such cases, may one argue that the police should intervene to commit those so unstable as to constitute a danger to others and to arrest those who decide to act in ways that endanger others, but not those who simply distribute ideas and images with non-malevolent intent and in circumstances not analogous to the shouting of "Fire!" in a crowded theater? The argument in favor of placing the burden on those who need to be protected against uncontrollable passions of fanaticism depends on the circumstance that ideas are usually not causes of action, but reasons for actions, and that punishment should begin and end with the persons who commit the evil acts. Of course, if one spreads an idea with evil intent, then one may in principle be punished as with laws against conspiracies. One will have joined in the evil action. In short, competent maleficent fanatics should be punished and incompetent dangerous fanatics can be made subject to restraining force or, in certain circumstances, unconsented-to treatment.

If one carries these considerations over to the issues of suicide and euthanasia, it follows that the burdens should be placed on the shoulders of those who might incompetently commit suicide or engage in involuntary euthanasia, were easy access available. It is such incompetents who should be treated and restrained: they may be constrained when it involves an area of their lives where there is no freedom to respect or restrict. However, those who can choose competently should remain unhindered. The slippery slope argument may lead to greater therapeutic vigilance and to the development of more efficient means of punishing those who act immorally, but not to restricting the actions of peaceable individuals with themselves and consenting others.

6. Maintaining the ethical integrity of the medical profession

It has been argued that individuals should not be allowed to refuse treatment or expedite their deaths while under medical care or while within a hospital, for if such were tolerated, it would undermine the ethical integrity of the medical profession. In addition, many have the sense that physicians should not be involved in aiding the deaths of patients. Such hesitations have often been attributed to the Hippocratic Oath, a document not authored by Hippocrates and appealing in part for its justification to the Greek gods and goddesses. If a monotheist can amend the Oath to leave out pagan piety, or a modernist to leave out commitment to the financial support of one's teachers, it would also seem possible that one could reinterpret the Hippocratic proscriptions of physicians' facilitating active euthanasia in order to disallow only actions without the consent of patients. An example can be found in the recent resolution of the Royal Netherlands Medical Association [30].

If authority is not available to impose a particular moral view of the good life through state force, then the "orthodox" medical profession will need to rely on a demonstration of its virtues in order to gain patients' compliance with its views of the good death. In addition, those who wish to benefit from voluntary euthanasia should not hope to do so in a Mount Sinai or Saint Elizabeth's hospital. Religiously affiliated hospitals may set policy so that anyone who is admitted to the hospital or employed by the hospital is warned that such a relationship requires compliance with the hospital's moral standards. In such circumstances the authority of the hospital can be clearly derived. The moral integrity of, for example, Catholic or Jewish physicians or nurses has a concrete meaning. Members of religiously oriented associations of physicians and nurses should not be compelled to violate their consciences. However, if such physicians and nurses decide to work in a secular hospital, in doing so they may have already compromised elements of their moral commitment. One might even imagine the extreme case of a Seneca Memorial Hospital that offers suicide and euthanasia. One cannot accept employment there and then protest that one's moral integrity is being violated if asked to assist in a suicide or voluntary euthanasia.

7. Parens patriae, paternalism, and protecting peoples' best interests, whether or not they want it

The foregoing reflections on the limits of state authority cast into doubt the right of the state to use force to impose on competent individuals the state's view of what is in their best interests. The more one is skeptical concerning the ability of others to determine the correct ranking of harms and benefits and the more one is skeptical about the authority to impose by force such a ranking, were it discoverable, the more individuals are free to make their own determinations of their own best interests and to live by the consequences. The purchase price of taking freedom seriously is tragedy.

Such considerations against state-imposed strong paternalism do not bear on weak paternalism, where the state, physicians or others may intervene when it is unclear whether individuals are choosing competently or with sufficient information. Once competence is established and information offered, the authority to intervene on the presumption that such an intervention would be accepted vanishes. Arguments such as Gerald Dworkin's [10] regarding the state having an implicit contract to intervene paternalistically are narrowed. Such arguments may be able to establish the propriety of weak paternalistic interventions. The authority for such interventions ceases as soon as it is established that the recipient of the intervention competently rejects the intervention.

These considerations do not undermine the authority to intervene for treatment and restraint on the basis of past competent instructions. One can create a Ulysses agreement with one's physician, psychologist, or others, authorizing intervention to prevent suicide, even when competent. Here the authority to intervene is clear. One might here also imagine that accepting a commission in the Armed Forces would involve ceding the right to commit suicide, except under specified circumstances that would not compromise battle-readiness. So, too, the acceptance of an unsecured loan might involve renouncing the right of suicide until the loan is repaid or security provided.

8. Protecting innocent third parties

It is implausible that general social relations in and of themselves create obligations that defeat the right of suicide. But insofar as individuals can divorce and/or put their children up for adoption, individuals, with proper provision for the care of those to whom they have obligations, may commit suicide or make use of euthanasia. Further, if death is imminent, little if anything may be denied to those to whom one has obligations, if one accelerates one's death by a few days, weeks, or months. As Hume reflected, "... suppose that it is no longer in my power to promote the interest of society; suppose that I am a burthen to it; ... In such cases my resignation of life must not only be innocent but laudable" ([21], p. 413). Since insurance policies often pay after a specific period despite suicide, it is possible for individuals to make provision for dependents, should suicide ever need to be embraced.

In the narrow sense of protecting innocents, the state has authority to intervene. It may and should come to the aid of those who are about to be made the subject of involuntary euthanasia. It may as well come to the aid of those whose contractual rights would be violated, should an individual commit suicide or make use of voluntary euthanasia. However, such interventions require showing that actual rights are in fact being violated. Vague claims of obligations are not sufficient to justify coercive force. Again, it is the force user who bears the burden of justification.

9. The state: coming to terms with free men and women

The analysis in section II established the limits of state authority on the basis of limits of reason to establish a morally canonical view of the good death and/or the authority to impose it on unconsenting incompetents who have not through special contract ceded their right to determine the circumstances of their own deaths. If one cannot establish that the king, the state, or majorities rule by divine right, or if one cannot establish a morally canonical view of the good life, which may be imposed by force, or if people have not entered into special relationships that preclude them from ending their lives by themselves or through the aid of consenting others, then the authority of the state to intervene in such choices is dramatically circumscribed. The rights to privacy in the choice of the circumstances of one's death disclose the limits of the established authority of others to intervene. In this section these general reflections and their conclusions have been applied to the various traditional ways in which state intervention has been justified in forbidding suicide or voluntary euthanasia. The argument of this essay has not been a positive argument in the sense of a contrary contention in favor of a particular view of correct conduct. Rather, the consequences of taking seriously the limits of the capacity of reason to establish the authority to coerce innocent third parties have been applied to a series of considerations usually advanced to justify coercive societal force.

There still may be suggestive arguments that indicate that certain ways of living and dying are better than others. Or grace may bring the conviction of a particular religious understanding. It is in terms of such concrete views of the good life and of the good death that we know how to live and die. Different communities in different ways give instruction regarding the sanctity of life, courage, and freedom. In terms of such communities, we learn the value of dying as Seneca, as Akiba, as S . Laurence, as St. Olaf, as Torii Suneemon, as James Bonham, or as Yukio Mishima.

Considerable restraint is required in order to be willing simply to convert through witness and not to employ as well the coercive power of the state to make sure that others live in accord with our concrete moral visions. Still, such restraint is required by the minimal notion of the peaceable community. As the rationales for state intervention weaken, the state becomes ever more amoral with respect to particular visions of the good life. The state becomes a vehicle for protecting the peaceable choices of individuals, including choices of competent individuals to do things with themselves and consenting others what many, if not most, may hold to be immoral. Given centuries of expansive views of state authority, it is difficult for many to accept the limitations of our circumstances. Many appear to be of the opinion that if one really believes in something, if one really has moral commitments, then one should be willing to lay down the freedom and lives of others for these commitments. Such zeal notwithstanding, the difficulty is to demonstrate in terms of general rational ethics that such deportment is not simply a special instance of the use of unauthorized force against the unconsenting innocent.

IV. TEXAS AND HOLLAND: LIGHTS TO THE WORLD

1. Remaining hesitations

Despite all of the foregoing reflections regarding the limits of state power, governments seem generally quite willing to intervene in determining when individuals may quit this life. Two recent cases can be advanced as examples. One involves Elizabeth Bouvia, a severely handicapped woman suffering from cerebral palsy, who sought a permanent injunction against a hospital that wished to feed her without her consent. The court denied her request, arguing, *inter alia,* that providing her with a restraining order would "have a devastating effect on other patients within Riverside General Hospital and other physically handicapped persons who are similarly situated in this nation" ([4], p. 1243). The court concluded that "she does have the right to terminate her existence but not while she is non-terminal with the assistance of society" ([4], p. 1247).

One should note that the court's refusal to grant an injunction was not predicated only on a concern to prevent suicide ([4]), p. 1245), but a concern also to protect the hospital and medical profession from being involved in actions held to be immoral. The second set of considerations, as has been noted, is quite proper. Just as individuals have a right, *ceteris paribus,* by themselves or with the help of consenting others to leave this life, they do not have a right to call on unwilling hospitals and physicians, much less society, for aid. The same considerations that limit the authority of the state limit the right of individuals to involve others in their lives and deaths. Despite some very troublesome elements of the opinion of the Superior Court of the State of California in the County of Riverside, there are wholesome elements as well.

The same can be said with regard to the Probate and Family Court opinion in the Brophy case [7]. This case involved Paul E. Brophy, a former emergency medical technician, who at the age of 45 suffered a subarachnoid hemorrhage as a result of a ruptured basilar tip aneurysm on March 22, 1983. This malevent left Brophy's body in a permanently vegetative state. He had previously indicated that he would not wish to have his life sustained, should he permanently lose consciousness. In response to a request by Mrs. Brophy to stop all artificial feeding, the Court of the first instance held that

[t]here are circumstances in which the fundamental right to refuse extremely intrusive treatment must be subordinated to various State interests. Among the State interests which have been identified in prior cases: (1) the preservation of life; (2) the protection of the interests of innocent third parties: (3) the prevention of suicide; (4) maintaining the ethical integrity of the medical profession ([7], p. 39).

Again, the court was correct that hospitals and medical professionals have a right not to become involved in undertakings they hold to be immoral.

The first Brophy court went further. It was willing to use state force, even if innocent third parties or unwilling physicians were not involved. The court foreclosed any possibility of Mrs. Brophy's following her husband's wishes by specifying that

[i]n the event that Patricia E. Brophy, guardian of Paul E. Brophy, causes the ward to be transferred to another medical facility, she is permanently enjoined from authorizing said facility to either remove or clamp Paul E. Brophy's gastrostomy tube for the purpose of denying said ward hydration and nutrition required to sustain his life [7].

The first court took this position despite the fact that it recognized that Mr. Brophy, on the basis of the statements he made while competent, would have rejected such treatment being provided to him ([7], p. 25).

Governments generally remain unwilling to allow competent individuals to dispose of their own lives. This unwillingness is perhaps most perturbing in cases such as Brophy (subsequently the first decision was overruled and Brophy has died [17]), where the life being preserved is mere vegetative life, and therefore of no plausible benefit to the person who once lived in the body. It is likely that the reticence to allow people to choose freely springs both (1) from an unwillingness to allow people to embrace moral positions radically different from the controlling societal majority, and (2) from the fear that freedom will lead to untoward consequences. There is little to be said here regarding the first issue. Tyrants should be set aside. With regard to the second, whether it is possible safely to allow people to make their own life and death choices, the answer is reminiscent of the one offered by H.L. Mencken, when asked if he believed in infant baptism ("Hell, yeah, I've seen it done"). As has already been indicated, Texas until very recently did not proscribe suicide or aiding and abetting suicide [11]. In addition, the Kingdom of the Netherlands has now moved informallv to allow euthanasia on request ([20], [30]). Significant abuses and untoward consequences are hard to document. It appears not only morally unavoidable but practically unproblematic to allow competent individuals to choose their exit from this life.

There is evidence of change in the American law. The California Court of

Appeals in reviewing the Bouvia case reserved the earlier court ruling and held that,

> [o]verlooking the fact that a desire to terminate one's life is probably the ultimate exercise of one's right to privacy, we find no substantial evidence to support the court's conclusion. Even if petitioner had the specific intent to commit suicide in 1983, while at Riverside, she did not carry out that plan. Then she apparently had the ability, without artificial aids, to consume sufficient nutrients to sustain herself, now she does not. That is to say, the trial court here made the following express finding, "Plaintiff, when she chooses, can orally ingest food by masticating 'finger food' *though additional nutritional intake is required intravenously and by nasogastric tube.*" ... (Emphasis added). As a consequence of her changed condition, it is clear she has now merely resigned herself to accept an earlier death, if necessary, rather than live by feedings forced on her by means of a nasogastric tube. Her decision to allow nature to take its course is not equivalent to an election to commit suicide with real parties aiding and abetting therein. (*Bartling v. Superior Court, supra,* 163 Cal.App.3d 1986; *Lane v. Candura, supra,* 376 N.E.2d 1232).

Moreover, the trial court seriously erred by basing its decision on the "motives" behind Elizabeth Bouvia's decision to exercise her rights.

> If a right exists, it matters not what 'motivates' its exercise. We find nothing in the law to suggest the right to refuse medical treatment may be exercised only if the patient's *motives* meet someone's else approval. It certainly is not illegal or immoral to prefer a natural, albeit sooner, death than a drugged life attached to a mechanical device. It is not necessary here to define or dwell at length on what constitutes suicide. Our Supreme Court dealt with the matter in the case of *In re Joseph G.* (1983) 34 Cal.3d 429, wherein declaring that the State has an interest in preserving and recognizing the sanctity of life, it observed that it is a crime to aid in suicide. But it is significant that the instances and the means there discussed all involved affirmative, assertive, proximate, direct conduct such as furnishing a gun, poison, knife, or other instrumentally or usable means by which another could physically and immediately inflict some death producing injury upon himself. Such situations are far different than the mere presence of a doctor during the exercise of his patient's constitutional rights ([5], pp. 23–25).

The court, in short, allowed individuals to commit what would count for many as the mortal sin of suicide. The court also affirmed strong rights of privacy, which include the right to refuse life-saving treatment even in circumstances when the patient is not terminal or permanently comatose.

The concurring opinion of Justice Compton touches on even more fundamental issues. It goes beyond the views developed in early Texas law and reaches towards the current policy in the Kingdom of the Netherlands.

> The right to die is an integral part of our right to control our own destinies so long as the rights of others are not affected. That right should, in my opinion, include the

> ability to enlist assistance from others, including the medical profession, in making death as painless and quick as possible.
>
> That ability should not be hampered by the state's threat to impose penal sanctions on those who might be disposed to lend assistance.
>
> The medical profession, freed of the threat of governmental or legal reprisal, would, I am sure, have no difficulty in accommodating an individual in Elizabeth's situation ([5], pp. 2–3).

Whether American courts will adopt Compton's reasoning, only the future can tell. If the foregoing arguments hold, they have no moral choice but to do so. The experience with both the early history of Texas and currently in the Kingdom of the Netherlands may help to assure individuals that widespread horrors will not occur if the rights of competent individuals are respected with regard to suicide, assisting suicide and voluntary euthanasia.

2. How bad would it be?

Left to their own devices, free men and women would undoubtedly at times commit rational suicide. Often in such circumstances, they would ask for the assistance of third parties. More frequently, they would likely seek the provision of voluntary euthanasia, especially when faced with periods of protracted debility and/or suffering. In addition, individuals would likely establish advance directives for voluntary euthanasia in the event that they would become severely obtunded or senile. Procedural safeguards would need to be established to ensure that competent wishes were faithfully recorded and that individuals could easily revoke directives once enacted. One would need to face the question of whether competent requests for euthanasia could be voided by subsequent dubiously competent revocations. Euthanasia directives would in principle raise few issues not already associated with the advance directives currently allowed in the United States. But there would be differences. Unlike present directives, one would be clearly recognizing individual rights in circumstances other than those of terminal illness. One would also not be forcing individuals to exit this life only through the slow deterioration of functions, once "artificial means of life support" had been withdrawn. Once the decision to exit had been competently made, it could lead to an expedited departure from this world.

Would people always choose wisely? Surely not. Would it be important to educate people concerning how to make prudent choices? Most assuredly it would. Would more people go to hell? God only knows. Would there be less suffering to release the poor souls in purgatory? Perhaps. Should priests suggest that one suffer to the end in order both to save one's soul and to shorten one's time in purgatory? I am not sure. Instead, the narrow question in this essay has been whether coercive state force may be used to prevent competent individuals from choosing suicide, assisting in suicide, or providing or making use of voluntary euthanasia. The answer appears to be in the negative. Honest men and women should therefore come to terms with the ways in which they must live with the free choices of individuals whose views of the good life and good death diverge radically from theirs.

NOTES

[1] In this paper I will use suicide to mean death brought about by willful omission or commission for the purpose of ending life. Under aiding and abetting suicide I will include all intentional encouragement of suicide, whether by words or by the provision of material. Under euthanasia I will include all intentional acts or omissions for the purpose of bringing about another's death in order to assist that person in avoiding suffering and pain, whether associated with living or with the process of dying itself. Euthanasia will be termed voluntary euthanasia when it occurs at the behest of the person killed, either directly or indirectly, as through an advance directive. This broad definition of euthanasia includes acts that some might term active assistance of the suicide of another (e.g., shooting someone at his request), in that it is not restricted simply to easing the process of dying. I restrict the compass of aiding and abetting suicide and expand that of euthanasia in order to use a terminology congruent with that of the Texas Court of Criminal Appeals in Grace v. State, 44 Tex. Crim. 193, 69 S.W. 529 (1902) and Sanders v. State, 54 Tex. Crim. 101, 105, 112, S.W. 68, 70 (1908). For the Texas Court, aiding and abetting suicide, which it allowed, included only encouragement by word or by the provision of material; the suicide, however, had to perform the last act himself. Unfortunately, Texas in 1973 criminalized aiding and abetting suicide so that it is now a felony; Texas Penal Code Annotated, s22.08 (Vernon 1974). In this change, one finds one of many examples of the untoward influences on the moral fabric of the state due to excessive immigration from the Northern states. This way of speaking of euthanasia is also in some respects restrictive, in that it does not count as euthanasia those actions or omissions foreseeably leading to an easier death, as long as that death is not intended. Duelling will usually not be a case of either suicide or euthanasia, unless one of the parties intends not to win and survive and deports himself in a way likely to lead to losing the duel, or if one duels with another in order to bring him an easier death or achieve some goal such as usually sought through euthanasia.

Distinctions between active and passive euthanasia, between killing and letting die, will be important primarily in terms of the consequences associated with these two different ways of bringing about another's death. Passive euthanasia can often be more easily described falsely as a foreseen but not intended death of another, and therefore not acknowledged as a case of euthanasia, as I have defined the term. Such disingenuity is justified at times on the basis that unacknowledged euthanasia is emotionally less distressing. In addition, some hold that openly allowing active euthanasia would undermine restraints against taking of life without consent.

[2] One of the difficulties is to decide how sure one ought to be of the incompetence of an individual before one uses force in order to constrain or to provide treatment without consent. See [6].

[3] Rationality with regard to autonomy of action must be judged in terms of the capacity of individuals to understand and appreciate the consequences of their actions and not their willingness to embrace a particular set of premises held by some to be the correct premises. Rationality must be understood in terms of process, not content, because the goal is to identify those individuals who can be blamed and praised for their actions, not those who are acting correctly. Innocent individuals who can be blamed and praised for their actions may not be used without their consent, not because they are acting correctly, but because such use would be a willful rejection of the possibility of mutual respect [11].

[4] Permanently unconscious but not yet whole-brain-dead human bodies have a peculiar legal standing. Their relatives can sue for funds sufficient to sustain them *ad indefinitum*. When such suits are successful, the award usually ranges in the millions of dollars. As it is, after a lump-sum, non-structured settlement, the family can subsequently request do-not-resuscitate orders to be written and/or take other steps to have treatment stopped so as to inherit the funds set aside to sustain the body. As the President's Commission stated, "The decisions of patients' families should determine what sort of medical care permanently unconscious patients receive. Other than requiring appropriate decisionmaking procedures for these patients, the law does not and should not require any particular therapies to be applied or continued, with the exception of basic nursing care that is needed to ensure dignified and respectful treatment of the patient" ([29], p. 6). There are good reasons to conclude that such bodies no longer are the bodies of living persons ([12], pp. 214–16), and, among other things, that courts should not award funds for the long-term support of such bodies. In addition, given a more sensible assessment of the status of such bodies, it would be recognized that they cannot be the objects of acts of murder or euthanasia. There is no one left to kill.

[5] Individuals are likely to take into consideration not only suffering in the sense of an adverse appreciation of pain, but also the psychological and social costs that their further life will likely impose on those for whom they care and on society generally. The costly prolongation of life may expend family resources, which the individual would rather have invested in other undertakings (e.g., sending children to college). Finally, the person may find continued existence under certain circumstances to be undignified and unacceptable from an aesthetic point of view.

[6] The doctrine of double effect in Catholic moral theology allows individuals to engage in acts or omissions that might lead to a quicker death, as long as (1) the actions themselves are not intrinsically immoral, (2) the goods sought (e.g., relieving pain or decreasing costs) are not achieved through the non-morally-evil outcome (e.g., death) due to the omission or commission, (3) one does not will the evil outcome (e.g., the death of the patient was simply foreseen but not intended), and (4) there is a proportionate good (e.g., relief of severe pain or the avoidance of disproportionate costs). For a study of the doctrine of double effect and the difference between

foresight and understanding, see [24].

[7] Though Moscow may in some respects be regarded as a latter-day Athens (one might think of how oppressive the government of Plato's *Laws* would have been), one must recognize that Marxism has many analogies to the Catholic Church in the Middle Ages. It is replete with dogma, hierarchy, and inquisition, though on the surface it claims that its tenets are scientifically supportable. The Vatican makes similar claims, given the Church's strong assertions regarding natural theology and the capacities of reason. The civilizations of Europe in general show the results of a mixed marriage between Jerusalem and Athens.

[8] By freedom as a side constraint, I mean the condition of respecting individuals in the sense of asking their permission before things are done with them. By freedom as a value, I mean imputing a worth or ranking to liberty in relation to other goods. In terms of freedom as a side constraint, one asks recruits permission before one makes them members of the French Foreign Legion. In terms of freedom as a value, one does not join the French Foreign Legion unless it is necessary as a condition to maintain one's liberties. See, for example, [27], pp. 30–34.

[9] For a review of the evolution of Texas law regarding suicide, see [11]. One should note that duelling was not proscribed until December, 1836. See [19], p. 288. In this vein, one should not legally condemn gladiatorial combat when all participants are willing or have been duly sentenced, as occured under Emperor Honorius, after Telemachus's dramatic protest in the Coliseum in 404.

[10] Of course, there never has been nor will there be a golden age, even for libertarians. However, Iceland and early Texas provide examples of states with very minimal governments. For a study of Icelandic legal and social systems, see [15]. The reader may also wish to consult the Njal Saga and the Gragas, which is the earliest compilation of Icelandic law. David Friedman has provided a study of the efficiency of the private enforcement of justice in Iceland, [16], [17].

[11] Controversies can come to a conclusion through an appeal to: (1) force, (2) common consensus, (3) sound moral argument, or (4) common negotiation and agreement. The first mode of controversy resolution gives no intellectual satisfaction, though it may erase the external signs of the dispute. In the second case, there is also no intellectual satisfaction: Resolution is the result of a sociopsychological fact. Moreover, consensus is unlikely to obtain in any complex society where there is a plurality of viewpoints. In peaceable pluralist societies that are secular in the sense of not embracing a particular orthodoxy, controversies will need to be resolved either through rational argument or common agreement. In the first case one will be seeking a correct solution, in the second case one will be seeking a fair solution. See [13]. Tom Beauchamp explores a number of these themes, though he provides a slightly different characterization of the modes for resolving controversies [1].

[12] I use deontological to identify those moral claims that are argued to hold independently of considerations of their consequences. In contrast, I use teleological to identify those moral claims that are held to depend on their consequences. Deontological claims identify what is right to do, independent of what is useful to do.

[13] Up to three states, Alabama, Oregon, and South Carolina, may still have a basis at law for holding suicide to be a crime. See Southern Life & Health Ins. Co. v. Wynn., 29 Ala. App. 209, 194 So. 421 (1940); Wyckoff v. Mutual Life Ins. Co., 173 Dr. 592, 147 P.2d 227 (1944); and State v. Levell, 13 S.E. 319 (S.C. 1891). Attempted suicide

may still be regarded as a crime in Illinois and Indiana. See Royal Circle v. Achterrach, 204 Ill. 549, 68 N.E. 492 (1903), and Wallace v. State, 232 Ind. 700, 116 N.E.2d 100 (1953). In addition, Massachusetts has held that attempted suicide is an unpunishable crime; Commonwealth v. Mink, 123 Mass. 422 (1877), and Commonwealth v. Dennis, 105 Mass. 162 (1870).

[14] For the edification of the reader, let me take note of one of the first, if not the first, executions of a person on the North American continent for a crime against nature. The sentence was carried out on one Benjamin Goad. He was hanged and his mare executed in his sight after he had been discovered in 1673 committing the "unnatural & horrid act of Bestillitie on a mare in the highway or field". *Records of the Court of Assistance of the Colony of the Massachusetts Bay 1630–1682 (1901)*, vol. 1, pp. 10–12.

[15] If the state does not own its subjects, and if there are no general moral arguments to establish that suicide is wrong and that others have a right to use force in preventing that wrong, then in the absence of special contracts to the contrary, *volenti non fit injuria*. As has been noted above, this appears to have been the position of early Texas law, which did not forbid suicide, aiding or abetting suicide, or duelling. However, the view that consent cures has not been generally accepted in Anglo-American law, the inspirations of the early Republic of Texas to the contrary notwithstanding. Consider the early contrary case of Matthew v. Ollerton, where the court held that, "If a man license Another to beat him, such license is void as it is against the peace". Roger Comberbach, Matthew versus Ollerton (1693), in *The Report of Several Cases Argued and Adjudged in the Court of King's Bench at Westminster* (London: J. Walthoe, 1724), p. 218.

[16] The highest appellate court of Massachusetts reversed the first decision and sustained the request to discontinue all life prolonging treatment including artificial hydration and nutrition in *Brophy v. New England Sinai Hospital, No. 4152* (Mass. Supr. Ct., Sept. 11, 1986).

BIBLIOGRAPHY

[1] Beauchamp, T.: 1986, 'Ethical Theory and the Problem of Closure', in H.T. Engelhardt and A. Caplan (eds.), *Scientific Controversies*, Cambridge University Press, New York, pp. 27–48.

[2] Berenson, R.A.: 1984, *Intensive Care Units (ICUs): Clinical Outcomes, Cost and Decision Making*, Office of Technology Assessment, Washington, D.C.

[3] Blackstone, W.: 1783, *Commentaries on the Laws of England*, 9th ed., W. Strahan, London.

[4] Bouvia v. Riverside General Hospital, No. 159780 (Cal. Super. Ct., Dec. 16, 1983).

[5] Bouvia v. L.A. Co. Sup. Ct., 2nd Civ. No. B019134 (April 16, 1986), p. 8.

[6] Brody, B. and Engelhardt, H. (eds.): 1980, *Mental Illness: Law and Public Policy*, D. Reidel, Dordrecht, Holland.

[7] Brophy v. New England Sinai Hospital, No. 85Eooo9–GI (Mass. Trial Ct., Oct. 21, 1985).

[8] Burnett v. People, 204 Ill. 208, 68 N.E. 505, 510 (1903).

[9] Camus, A.: 1961, *The Myth of Sisyphus and Other Essays*, trans. by J. O'Brien, Alfred A. Knopf, New York.

[10] Dworkin, G.: 1972, 'Paternalism', *Monist* 56, 64–84.
[11] Engelhardt, H.T., Jr., and Malloy, M.: 1982, 'Suicide and Assisting Suicide: A Critique of Legal Sanctions', *Southwestern Law Journal* **36,** 1003–37.
[12] Engelhardt, H.T., Jr.: 1986, *The Foundations of Bioethics,* Oxford University Press, New York.
[13] Engelhardt, H.T., Jr., and Caplan, A. (eds.): 1986, *Scientific Controversies,* Cambridge University Press, New York.
[14] Engelhardt, H.T., Jr.: 'Texas: The Meaning of its Myths and the Message from its Heroes', to be published.
[15] Foote, P.G. and Wilson, D.M.: 1970, *The Viking Achievement,* Sidgwick & Jackson, London.
[16] Friedman, D.: 1984, 'Efficient Institutions for the Private Enforcement of Law', *The Journal of Legal Studies* **13,** 379–97.
[17] Friedman, D.: 1979, 'Private Creation and Enforcement of Law: A Historical Case', *The Journal of Legal Studies* **8,** 399–415.
[18] Hales v. Petit, 1 Plowden 253, 76 Eng. Rep. 387 (1562).
[19] Hartley, O.C. (ed.): 1850, *Digest of the Laws of Texas,* Thomas Cowperthwait, Philadelphia.
[20] Hoge Raad der Nederlanden, Strafkamer nr. 77.091, November 27, 1984.
[21] Hume, D.: 1964, 'Of Suicide', in *Essays Moral, Political and Literary,* ed. T.H. Green and T.H. Grose, Scientia Verlag Aalen, London.
[22] Kant, I.: 1909, *Fundamental Principles of the Metaphysic of Morals,* in *Critique of Practical Reason and Other Works on the Theory of Ethics,* trans. by T.K. Abbott, 6th ed., Longmans, Green and Co., London.
[23] Kant, I.: 1964, *The Metaphysical Principles of Virtue: Part II of the Metaphysics of Morals,* trans. by J. Ellington, Bobbs-Merrill, Indianapolis, Indiana.
[24] McCormick, R. and Ramsey, P. (eds.): 1978, *Doing Evil to Achieve Good,* Loyola University Press, Chicago.
[25] Nathan, J.: 1974, *Mishima: A Biography,* Little and Brown, Boston.
[26] Nietzsche, F.: 1954, *Thus Spake Zarathustra,* in *The Philosophy of Nietzsche,* Modern Library, New York.
[27] Nozick, R.: 1974, *Anarchy, State and Utopia,* Basic Books, New York.
[28] People v. Onofre, 415 N.E.2d at 940 n.3, 434 N.Y.S.2d at 951 n.3.
[29] President's Commission for the Study of Ethical Problems in Medicine and Biomedical and Behavioral Research: 1983, *Deciding to Forego Life-Sustaining Treatment,* U.S. Government Printing Office, Washington, D.C.
[30] Royal Netherlands Medical Association: 1984, 'Standpunt inzake euthanasie', *Medisch Contact* **39.**
[31] Schenck v. United States, 249 U.S. 47 (1919).
[32] Seneca: 1958, 'Letter on Suicide', in *The Stoic Philosophy of Seneca,* trans. by M. Hadas, Norton, New York, pp. 204–205.
[33] Tacitus: 1942, *The Complete Works of Tacitus,* trans. by M. Hadas, Random House, New York.

Baylor College of Medicine,
Houston, Texas, U.S.A.

NOTES ON CONTRIBUTORS

Darrel W. Amundsen, Ph.D., is Professor of Classics, Western Washington University, Bellingham, Washington.

Tom L. Beauchamp, Ph.D., is Professor of Philosophy, and Senior Research Scholar, Kennedy Institute of Ethics, Georgetown University, Washington, D.C.

Joseph Boyle, Ph.D., is Professor of Philosophy, St. Michael's College, University of Toronto, Toronto, Ontario, Canada.

Baruch A. Brody, Ph.D., is Leon Jaworski Professor of Biomedical Ethics and Director of the Center for Ethics, Medicine, and Public Issues, Baylor College of Medicine, and Professor of Philosophy, Rice University, Houston, Texas.

John M. Cooper, Ph.D., is Professor and Chairman, Department of Philosophy, Princeton University, Princeton, New Jersey.

H. Tristram Engelhardt, Jr., Ph.D., M.D., is Professor of Medicine and of Community Medicine, and Member, Center for Ethics, Medicine, and Public Issues, Baylor College of Medicine, Houston, Texas.

Gary B. Ferngren, Ph.D., is Professor, Department of History, Oregon State University, Corvallis, Oregon.

INDEX

The Philosophy and Medicine Book Series

Editors

H. Tristram Engelhardt, Jr. and Stuart F. Spicker

1. **Evaluation and Explanation in the Biomedical Sciences**
   1975, vi + 240 pp. ISBN 90–277–0553–4
2. **Philosophical Dimensions of the Neuro-Medical Sciences**
   1976, vi + 274 pp. ISBN 90–277–0672–7
3. **Philosophical Medical Ethics: Its Nature and Significance**
   1977, vi + 252 pp. ISBN 90–277–0772–3
4. **Mental Health: Philosophical Perspectives**
   1978, xxii + 302 pp. ISBN 90–277–0828–2
5. **Mental Illness: Law and Public Policy**
   1980, xvii + 254 pp. ISBN 90–277–1057–0
6. **Clinical Judgment: A Critical Appraisal**
   1979, xxvi + 278 pp. ISBN 90–277–0952–1
7. **Organism, Medicine, and Metaphysics**
   Essays in Honor of Hans Jonas on his 75th Birthday, May 10, 1978
   1978, xxvii + 330 pp. ISBN 90–277–0823–1
8. **Justice and Health Care**
   1981, xiv + 238 pp. ISBN 90–277–1207–7(HB) 90–277–1251–4(PB)
9. **The Law-Medicine Relation: A Philosophical Exploration**
   1981, xxx +•292 pp. ISBN 90–277–1217–4
10. **New Knowledge in the Biomedical Sciences**
    1982, xviii + 244 pp. ISBN 90–277–1319–7
11. **Beneficence and Health Care**
    1982, xvi + 264 pp. ISBN 90–277–1377–4
12. **Responsibility in Health Care**
    1982, xxiii + 285 pp. ISBN 90–277–1417–7
13. **Abortion and the Status of the Fetus**
    1983, xxxii + 349 pp. ISBN 90–277–1493–2
14. **The Clinical Encounter**
    1983, xvi + 309 pp. ISBN 90–277–1593–9
15. **Ethics and Mental Retardation**
    1984, xvi + 254 pp. ISBN 90–277–1630–7
16. **Health, Disease, and Causal Explanations in Medicine**
    1984, xxx + 250 pp. ISBN 90–277–1660–9
17. **Virtue and Medicine**
    Explorations in the Character of Medicine
    1985, xx + 363 pp. ISBN 90–277–1808–3
18. **Medical Ethics in Antiquity**
    Philosophical Perspectives on Abortion and Euthanasia
    1985, xxvi + 242 pp. ISBN 90–277–1825–3(HB) 90–277–1915–2(PB)
19. **Ethics and Critical Care Medicine**
    1985, xxii + 236 pp. ISBN 90–277–1820–2

20. **Theology and Bioethics**
Exploring the Foundations and Frontiers
1985, xxiv + 314 pp. ISBN 90–277–1857–1
21. **The Price of Health**
1986, xxx + 280 pp. ISBN 90–277–2285–4
22. **Sexuality and Medicine**
Volume I: Conceptual Roots
1987, xxxii + 271 pp. ISBN 90–277–2290–0(HB) 90–277–2386–9(PB)
23. **Sexuality and Medicine**
Volume II: Ethical Viewpoints in Transition
1987, xxxii + 279 pp. ISBN 1–55608–013–1(HB) 1–55608–016–6(PB)
24. **Euthanasia and the Newborn**
Conflicts Regarding Saving Lives
1987, xxx + 313 pp. ISBN 90–277–3299–4(HB) 1–55608–039–5(PB)
25. **Ethical Dimensions of Geriatric Care Value Conflicts for the 21st Century**
1987, xxxiv + 298 pp. ISBN 1–55608–027–1
26. **On the Nature of Health**
An Action-Theoretic Approach
1987, xviii + 204 pp. ISBN 1–55608–032–8
27. **The Contraceptive Ethos**
Reproductive Rights and Responsibilities
1987, xxiv + 254 pp. ISBN 1–55608–035–2
28. **The Use of Human Beings in Research**
With Special Reference to Clinical Trials
1988, xxii + 292 pp. ISBN 1–55608–043–3
29. **The Physician as Captain of the Ship**
A Critical Appraisal
1988, xvi + 254 pp. ISBN 1–55608–044–1
30. **Health Care Systems**
Moral Conflicts in European and American Public Policy
1988, xx + 368 pp. ISBN 1–55608–045X
31. **Death: Beyond Whole-Brain Criteria**
1988, x + 276 pp. ISBN 1–55608–053–0
32. **Moral Theory and Moral Judgments in Medical Ethics**
1988, vi + 232 pp. ISBN 1–55608–060–3
33. **Children and Health Care: Moral and Social Issues**
1989, xxiv + 349 pp. ISBN 1–55608–078–6
34. **Catholic Perspectives on Medical Morals**
Foundational Issues
1989, vi + 308 pp. ISBN 1–55608–083–2
35. **Suicicide and Euthanasia**
Historical and Contemporary Themes
1989, vi + 286 pp. ISBN 0–7923–0106–4